THE
FIBEROPTICS &
LASER
HANDBOOK

THE
FIBEROPTICS & LASER
HANDBOOK

BY EDWARD L. SAFFORD, JR.

TAB TAB BOOKS Inc.

BLUE RIDGE SUMMIT, PA. 17214

To all students both young and old who can imagine the growth of technology and who bring forth the new ideas and inventions that constantly make our world a better and nicer place to live.

FIRST EDITION

THIRD PRINTING

Printed in the United States of America

Library of Congress Cataloging in Publication Data

Safford, Edward L.
    The fiberoptics and laser handbook.

    Includes index.
    1. Fiber optics.   2. Lasers.   I. Title.
TA1800.S24   1984   621.36′92   83-18084
ISBN 0-8306-0671-8
ISBN 0-8306-1671-3 (pbk.)

Cover photograph courtesy of ITT Cannon Electric.

Contents

Effect—On the Dualism of Light Rays—The Concept of Light
Sources and Illumination—One Practical Robotic Application of
Optical Fibers

Perot Interferometer—A Laser Remote-Control System Concept—Modulating a Laser Beam with FM and AM Types of Modulation—Integrated Optical Circuits—Using the PIN Diode in a Laser Communications System—An Antenna for a 10.6 Micron Printed Circuit—Potential Uses of Lasers

Introduction

Thank you for examining this book! I hope and trust that it contains information that you will find useful, interesting, informative or all of these. I hope it will titillate your scientific curiosity and imagination. I hope that it will become a prized volume on your bookshelf.

This is an adventure into the world of *optics* and *lasers* and the world of electromagnetic radiations. We shall venture into the microcosm of light (said world described as small only because the wavelengths of light are small). Not discussed are the effects of light on our world and cosmos. There the world of light is truly gigantic—a macrocosm.

Fiberoptics has a delicate sound, and perhaps it conveys a small sense of the mysterious. We recognize the word relating to sight in the "optics," and somehow we think of something long and flexible when we think of the part of the word which spells "fiber." Both concepts are correct. We may also think of lenses and focusing and glasses and such in connection with "optics" and even *Huygen's* principle, and *Fermat's* principle, and *Young's* experiments, and coherence and noncoherence of light rays. We may recall *Michelson's* interferometer which is the basis for our current "New World" of gyroscopes using lasers and fiberoptics, and beam-splitting techniques. We may even think of some branches of fiberoptics in that they may be either glass fibers or silica fibers of the fused type that are conductive in the ultraviolet part of the light spectrum.

So we hint at the radiation qualities of light and from this inference we gain an intuitive insight into the fact that there may be differences in the fibers used in fiberoptics depending upon which part of the light spectrum will be transmitted through them. Perhaps we then will begin to wonder what the "mechanism" of light transmission actually is that causes this phenomena of containment to occur in these specially fabricated fibers. Where does the light come from? Where does it go? How is it controlled? These are some of the questions that now arise. In this adventure we will supply some of the answers to these questions.

In the *microcosm* sense, "precision" has a meaning that is perhaps more demanding and rigorous than its meaning in connection with other scientific phenomena in the *macrocosm*. Measurements must be much more precise and the use of instruments that can give this precision will be examined in some detail.

But, we don't want to lose sight of the trees for the forest, and so we begin this adventure in a somewhat elementary sense and at a level that we hope will enable you to proceed without difficulty. You may actually be far ahead of us and may have already taken this journey so that only a few "milestones" of recall will be necessary. If this be the case, then we invite your companionship on this adventure and would delight in closer communication via *SASE* in the future. Improvements, additions and "better-paths-to-travel" will always be one of our objectives. And so, we begin this adventure into the world of fiberoptics and laser technology—both are fascinating subjects!

Chapter 1
Some Required Background
Information and Concepts

Almost everyone who has heard or read or discussed fiberoptics knows that what is being talked about is very much like a radar waveguide. It is a flexible type of strand of some kind of glass that is able to channel light rays through it in a somewhat efficient manner.

The fiber-optics channel does not permit the desired light rays to escape from its interior. It is capable of handling two-way transmission of those light rays in such a manner that one can (but must not) look into one end and see what is happening at the other end. This is only true in some scientific and medical applications. Also, because there are ways of transmitting sound, perhaps even pictures, over these light rays, these fiberoptics transmission systems open up a completely new concept in communication systems which have unique advantages with regard to noise elimination and interference, as we shall presently analyze and explore.

That the light rays may be likened to radar waves might be deduced from a basic equation:

$$V = f\lambda$$ **Equation 1-1**

$V = 3 \times 10^8$ meters per second.
Also we find that

$$\frac{\text{Meters}}{\text{Second}} = \frac{\text{Hertz}}{\text{Second}} \times \frac{\text{Meters}}{\text{Hertz}}$$ **Equation 1-2**

and so, having "matched" units, we can then restate the equation in the form:

$$\lambda = \frac{V}{f}$$

Equation 1-3

(V = velocity of light)
(f = frequency in hertz)
(λ = wavelength in meters)

and we can let (f) approach infinity. This is, of course, the realization of what happens as we move out along the rf spectrum to and past the "light frequencies" region. We see that as (f) approaches infinity, the wavelength approaches zero, and supposedly that is the end of the spectrum. We use the word "supposedly" because there will or may be other emanations of some different kind than oscillations which occur beyond the oscillatory spectrum. Radar waves are frequencies below the "visible light" portion of the spectrum, but they *are* oscillations. Since we are extending that spectrum, then we assume light rays to be oscillating electromagnetic radiations also (but of a much shorter wavelength). Some scientists, however, refer to the "corpuscular" nature of light and energy parcels of *photons*.

If light rays are then electromagnetic in form, and if we can "guide" such radiations inside specially sized and shaped "waveguides", then why not use optical fiber as the guide? We know that radar waves "flow" through their metallic waveguides because of reflection from the walls and the peculiar manner in which they impinge upon the walls. Thus, it is easy to accept the concept that light rays may also be reflected in a similar manner, from a different type of substance more suitable to the reflection of light than a metallic wall might be. Light also can be caused to "flow" down the lengths of channeling forming an *optical-fiber* path.

This is the essence of fiberoptics. The channeling of light rays through paths of optical fibers and the generating of the proper frequency of light, and making it in a proper form (coherent as opposed to noncoherent) to permit easy passage through whatever type substance is being used for the channel. Note here that this implies that different types of substances (all of which fall into the "optical" or "glass" category) have some different characteristics that make the conductivity of certain light frequencies easier than other light frequencies. In an elementary concept it simply means some types of optical fibers conduct some light frequencies easier than they do other light frequencies. We must generate the proper light frequency to pass through a given type of fiber, just as we have to generate a certain radar frequency to pass easily through a waveguide of a certain dimension.

MODERN CONCEPTS OF LIGHT AND LIGHT PRODUCTION

We associate the creation of light with the passage of electricity through a wire in a glass bulb (our ordinary tungsten filament light bulb) or the passage of high potential electricity through such a gas as argon in a neon tube. We also know, from our elementary physics, that light is a component of the electromagnetic spectrum because *James Clerk Maxwell,* using *Gauss's* law for electricity and magnetism, and using *Ampere's* law in an extended form and *Faraday's* law of induction showed this to be true. He showed that the light rays were waves* which are electromagnetic in nature and he showed that all such rays or waves have the *same speed in free space.* That speed, as we already know, is 3×10^8 meters/sec. The range of electromagnetic radiations extends from a zero frequency (hertz/sec) to an infinite frequency, and those of which light is a section consist of frequencies whose wavelengths (λ) range from 10^{-2} through the ultraviolet region of (10^{-8}) on upward in frequency through the gamma-ray region of (10^{-14}) wavelength.

It should not be a surprise to us to consider that *if* the speed of light is a constant in free space, and if the speed consists of an average over an infinite distance, that there could and should be some "wiggles" in the speed over smaller finite distances and through various types of conducting media. We think of the reduction in the speed of propagation of light rays through various lenses, the increase or decrease of speed of propagation of radar-type waves through cavity resonators and waveguides. So long *as the total* speed component over an infinite distance in free space remains constant, we have not violated the law which Maxwell derived in his classic work. Experiments have shown these finite speed variations to be a reality and from these conditions many scientific developments have evolved such as light lenses and focusing and such. Since the days of Galileo, who was concerned with the "speed of light" question, values have been measured by various means. A value of (c) the speed of light in a microwave cavity has been determined to be 2.997925×10^8 meters/second, whereas *Galileo* found by his measuring means, a speed of light to be 2.997924×10^8 meters/sec.

---

*Einstein stated light is a "packet" or "quanta."

At this point in the text it will be appropriate to provide some definitions:

- ☐ Micron = (μ) = 10^{-6} meter
- ☐ Millimicron = (mμ) = 10^{-9} meter
- ☐ Angstrom = (Å) = 10^{-10} meter
- ☐ Visible light spectrum = approx. 450-675 (mμ) (to the human eye)
- ☐ Center of visible light spectrum approx. 550×10^{-7} mμ
- ☐ Reflected energy from absorptive surface:

$$p = \frac{\mu}{c} = \frac{\text{momentum energy}}{\text{speed of light}}$$

- ☐ Reflected energy from totally reflecting surface:

$$p = \frac{2\mu}{c} = \frac{\text{(2) momentum energy}}{\text{speed of light}}$$

As you note from the definitions, when we are concerned with the wavelengths of light, we are concerned with extremely small distances; the Angstrom being:

$$\frac{1}{10,000,000,000}$$

of a meter. You can calculate the length if you know that a meter is 39.370078 inches. You won't be able to imagine such a small distance which is representative of one wavelength of light. It is important to know that "light guides" such as light fibers or light conductors have to be designed in terms of wavelengths of the light frequency that they are to be able to channel. It is of interest to us to know that the orange krypton-86 light has the most sharply defined wavelength and so is used as a standard of measurement. Length measurements using a *green* light from a mercury-198 lamp is said to be accurate to one part in one billion! Our modern electronic instruments used to measure frequencies and wavelengths of light enable us to make and use such small values to the above accuracy!

The production of light is sometimes accomplished by heat. Witness the heat generated by an electric light bulb when electricity passes through its filament. It may be generated by lasers, as we know, when the atoms are stimulated and caused to go from one

4

energy level to another by the addition of energy to a synthetic ruby crystal or some kind of appropriate gas. We also know that the energy from a light bulb is noncoherent. That is, the waves strike out in a sperical or random manner from the filament much like the radio waves or microwaves emit from a suitable antenna for those frequencies. Lasers, on the other hand, emit *coherent* light rays. This means the wave fronts are all parallel. The frequency of the light being generated is a function of the type of laser that is the source of such light.

We can almost intuitively guess what we will need as a source for our light if we are going to send that light beam down some kind of special conducting fiber with as little loss as possible. We have said that some types of material conduct some frequencies of light better than they conduct other frequencies, so the frequency of the generated light is important to us. Also, at this moment we begin to suspect that whether or not the light is coherent may have some effect on its propagation through such a fiber-optics channel. We can begin to determine the correctness of our thinking if we next investigate light reflection and refraction, thinking about the way in which microwaves are reflected from the waveguide walls as this type energy passes down the waveguide channel. By the way, let us mention our book Modern RADAR (TAB Book number 1155) as a source for more information about radar frequencies and waveguides and related ideas.

LIGHT REFLECTIONS AND REFRACTIONS

If a light beam goes from one medium into another it is bent if the two materials have different properties of conductivity of the light rays. One classic example is that which takes place when a beam of light goes from the air into and through some water as illustrated in Fig. 1-1. Notice that the ray of light does not continue in the line of the air ray. It bends through some angle theta (θ) as shown. It has been shown in other works how a fish "sees" a human in a place where he is not physically present. Or how a human looking at a fish in the water sees that fish in a spot where he is not really located. Since our sight is based on the rays of light entering our eyes, and these rays showing the fish are reflected light rays, they emerge from the fish (and water) into another type medium of conductivity and so are bent.

It is important that we now consider how the ray is described as it approaches some different medium surface. The angle, measured from a perpendicular to that surface, is called the *angle of incidence* of the ray upon that surface. There is also a truth (axiom) in physics that light and other type rays will be *reflected* at an angle equal to the angle of incidence (provided that the angle is not so great that no reflection takes place). In Fig. 1-2 we show a section of rf waveguide and define these angles as they apply to the transmission of microwaves through that waveguide.

Immediately we are concerned that there might be a *critical angle* of incidence (such that to exceed this angle or not to exceed this angle) might result in a nonreflection (propagation) situation. If you have assumed or recalled that this is a true condition you are exactly correct. If you have studied the transmission of radiowaves, then you undoubtedly recall that when the waves "hit" the *Heaviside Layer,* some are reflected and some are not. Those which are not reflected "hit" this layer at less than the *critical angle* for reflection and so pass through the Heaviside layer along a refracted line.

THE LAWS OF REFLECTION AND REFRACTION

As in all good physics texts the laws for reflection and refraction can be stated as:

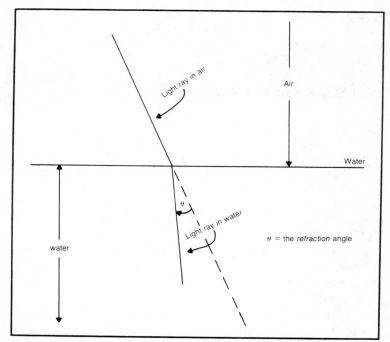

Fig. 1-1. The bending of a beam of light at the interface of air and water.

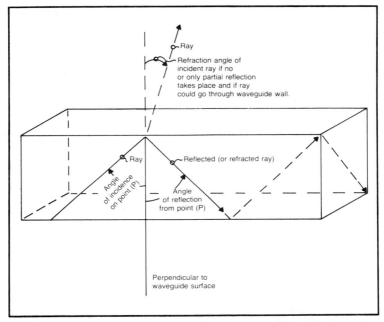

Fig. 1-2. Incidence, reflection, and refraction angles in a microwave waveguide.

(A) The reflected and refracted rays will lie in a plane containing the incident ray and the perpendicular (normal) to the surface of the material of reflection or refraction, and the angle of incidence will equal the angle of *reflection*.

(B) The equation: $\dfrac{\text{sin of angle of incidence}}{\text{sin of angle of } \textit{refraction}} = n_2/n_1$

where n_2/n_1 is a number called the *index of refraction* (a constant) of the second medium with respect to the first medium through which the ray passes.

The air-water index is about 1.33, *crown glass-air* is 1.52 and quartz is fused to have an index of 1.46. The air medium for these figures is assumed to be vacuum-like in composition with a refractive index of (1). We must also be aware that these index figures are for a given wavelength (5980 Å) and may change for other frequencies.

Thus, we refresh ourselves on some basic principles which will be important to us when we consider fiberoptics. In order to get a reflection we must have the juncture of two types of material such that there is some index of refraction between them, and we must not let the rays approach the *critical angle* or they will pass through and not be reflected at the junction. It is a minor note that prisms are based on the *refraction* of rays of light, and because different angles of refraction occur for different frequencies, we can use a prism to separate the light spectrum and "spread it out" on a surface so we can see it. It is also interesting that the laws of reflection and refraction can be derived from Maxwell's equations (if you want a mathematical challenge).

In our illustration (Fig. 1-2) we have indicated a rigid waveguide. In fiberoptics that use flexible fibers we will not necessarily have a rigid guide or channel. External pressures on the optical fiber may cause us some problems.

A SLIGHT BIT OF HISTORY

We know that light has been used as a medium by which some intelligence can be transmitted from one location to another. We recall that the early Indians used light reflections from mirrors to signal one-another that the wagon trains or soldiers were near. In modern warfare the signal light using a dot-dash or more properly, a "long-short" light blip code with *Morse* coding was used between ships to convey intelligence. The advantage is that the countermeasures against such a signaling system are rare and difficult to employ. Covert "eavesdropping" on such signals is difficult because of the directivity of the light beam, and yet the ability to convey intelligence of the type required is very high. Of course, the range is small as compared to that obtainable with radio waves due to the absorption and attenuation of the light energy by particles, moisture, and such in the atmosphere.

It is well known that in the 1880s Alexander Graham Bell spoke over a light beam. He used sunlight which was focused by means of a reflector (shades of the Indian days) and a lens to a device which could be made to vibrate in harmony with speech from a human vocal tract. The light beam was made to vary in-and-out of focus in such a manner that its strength upon a *selenium* detector was caused to vary. This, in turn, caused a varying electrical signal from the detector which could then be made to activate a telephone receiver in the usual manner and thus re-create the original speech. The distance between transmitter and receiver was very short compared to our modern capability of transmitting light beams. The principle involved was valid and so is still a basis for our current communication technology using light as the conducting medium. The method of varying the intensity (or *modulating* the light beam) is now different using solid-state technology and our detectors are improved but the basic concept is still the same.

It is known that the manner in which light beams are modulated to convey intelligence makes use of the principles of modulation found in modulating radio waves, and this results in a much more sophisticated system of intelligence transmission than was ever imagined by Alexander Graham Bell. Since the light spectrum is a continuation of the radio spectrum and since the frequencies increase and the wavelengths become shorter and shorter and since we know—this fact having been proved by scientists—that the information carrying capacity of a channel such as this will increase as the frequency of the channel increases, we now find ourselves wondering if we couldn't send a multitude of channels of information over one light beam. Of course, this is true. It has been proven by those skilled in such proofs that if one uses *only a small fraction of the bandwidth* available when using infrared or visible light frequencies, one such channel could carry and convey all the telephone conversations of every person on the North American Continent simultaneously!

To imagine in an elementary way how this might be done we need to recall the basic amplitude modulation equation:

$$AM = e = [A\,(1 + m\,e_s\,(t))]\sin(\omega_c t + \theta) \qquad \textbf{Equation 1-4}$$

Where
e = signal
A = amplitude constant
m = modulation factor less than unity
e_s = modulating signal
(t) = time
ω_c = carrier frequency in radians/sec.
θ = carrier angle of phase

and we know from the solution to this equation (or one of its type) that the intelligence is contained in the double sideband pair:

$$\omega_c - \omega_s \;;\; \omega_c + \omega_s \qquad \textbf{Equation 1-5}$$

ω_s = modulating frequency in radians/sec

Thus we know that when amplitude modulating a given carrier we actually transmit a band of frequencies that range from the frequency of the lower sideband to the upper sideband and are centered around the carrier frequency as shown in Fig. 1-3. It is nice

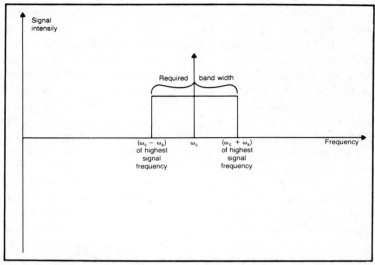

Fig. 1-3. Amplitude modulation (AM) with a radio carrier illustrating the bandwidth required.

that this illustration shows that a sizeable length of the frequency line is occupied with the band spectrum needed to transmit the intelligence contained in this signal. This "size relationship" is due to the fact that the frequency of the "carrier" (ω_c) is low (as radio frequencies are defined) and might be assumed to be a channel or carrier frequency in the broadcast band or a band *below* (having less hertz/sec) than found in the broadcast band.

Long ago our inventive telephone companies came up with a system of transmission of telephone messages called *"carrier telephone"*. Prior to this time we suppose that the telephone companies had to have *one pair* of wires for each telephone installed (unless of course several in a home or office were parallel wired to that single connecting pair of wires) and they determined that it was economically unsound to "string so many cables containing so many pairs of wires. This was true even if they used a *"common"* wire to all other wires to make up the telephone "pairs."

In the carrier telephone system, one pair of wires could be used to convey a broadband higher frequency signal and this could be broken down into bands that could be then modulated at one end by the transmitter and demodulated at the receiving end so that a multitude of telephone coversations were then "carried over a single pair of wires." It was much more economical with respect to both equipment, installation and maintenance and so consumer bills were less. The telephone company still made a good profit and everyone was happy. As the population grew, more of the problems encountered earlier began to reappear. Again the telephone company faced the situation and tackled the problem to try to find a solution. That solution will now come about through the use of light conducting fibers to replace the wires previously used. They can get so many bands in the light frequency range around one broadband light frequency that they can handle virtually millions of telephone conversations simultaneously over one channel.

This development was not without some problems. First they had to develop a suitable light source which could send an appropriate beam of light (at the desired frequency) over a light channel or fiber that was strong enough to be usable in a multitude of installations and over long distances. They had to invent some means of modulating that light beam in a most efficient manner and invent detectors which could reduce the light variations to suitably recognizable voices and other sounds.

Of course, you have already guessed that the laser—first invented and demonstrated in the 1960s—would furnish a light source. Also, LEDs (light-emitting diodes) were invented with sufficient brightness to be of use in this application. Neodymium-doped yttrium-aluminum-garnet (Nd:YAG) lasers emit a very narrow laser beam at about 1060 nm (nanometer) and this line-beam is easily sent through optical fibers of an appropriate type with low losses. LED type devices of the indium-gallium-arsenide phosphides operate at about 1200 to 1500 nm and when this frequency signal is sent through high-silica fibers there is even less loss than with the laser beam, and it is said that with LEDs on this wavelength, the material dispersion of glass becomes zero!

Detectors had to be developed, and one of these, the PIN diode had the fast response time to accommodate the modulation rate. The acronym PIN long worried yours truly, perhaps it does you. It stands for *p*-type silicon-*i*ntrinsic layer-*n*-type silicon sandwich! Now we know. Sometimes we think of an acronym as being an element or physical thing and never really learn what the abbreviation stands for, even though we use the element. It is of interest to us here to examine a sketch of the PIN photodiode in cross section in Fig. 1-4. This figure illustrates how the PIN diode works in reverse to that of a light source. The light rays that are absorbed produce the electron-hole pairs that result in the current flow. This current is directly proportional to the quantity and intensity of the light received.

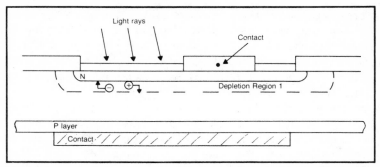

Fig. 1-4. A cross-sectional view of a PIN diode (courtesy of Northern Telecom).

We can examine a graph which illustrates the efficiency of some of the types of diodes (PIN photodiodes) made from silicon (the most common as of this writing) and germanium, as well as from some composite materials much written about in current literature; gallium (Ga) indium (In) and arsenic (As). It may be of interest to recall that germanium was often used in small solid-state radios as detectors "way-back-when." Perhaps you recall a coil of wire, an earphone and a germanium detector with a small capacitor by-pass across the "phones." The efficiency graph of the PIN diode types is illustrated in Fig. 1-5. This graph illustrates how efficiently a PIN diode made from the stated materials converts the light (at the various frequencies) to electricity.

Figure 1-5 shows that silicon PIN diodes work well at roughly 600 to 1000 nanometers (nm), while germanium and GaAs work well at the higher frequencies of 1600 to 1800 (nm). Notice that the composition photodiode made of GaAs has a nice linear output curve section at the lower frequencies with an efficiency of almost 50%. Quantum efficiency is defined as the fractional number of photons incident on the detector that are converted to signal current carrying electron-hole pairs. When the photon energy drops below the bandgap energy of the material the material becomes transparent to the light rays. The germanium-type detectors are said to have a rather large amount of thermal and "shot" noise. This is also true of avalanche (multiplying) type detectors. Silicon is said to have a low noise level. The avalanche photodiode acronym is (APD) and because of its low noise contribution to the current conversion, silicon is used in this application which results in a *gain* or multiplication of the light-to-current energy.

Bandwidth and recovery time is important in any electronic application and especially in the fiber-optics application. The speed of operation of a circuit and its bandwidth is governed to a very large extent by the RC time constants in the circuit (also the post detection filters used). In most *PIN* and *APD* diodes the limit is less than one nanosecond (ns) and this, in turn insures a bandwidth of at least 350 MHz that according to some companies is more than adequate for most of the foreseeable applications of these detectors. The construction of LEDs and lasers suitable for use in optical fiber applications is somewhat different as shown in Fig. 1-6.

Because a larger amount of light power from the laser source can be coupled effectively into optical fibers and because of the directivity of the light rays, these are preferred over the LEDs. Of course you would suspect that the type materials used would have been chosen for their efficiency in converting electrical energy into light energy. The material which forms the confinement areas must have a higher energy bandgap and a lower index of refraction than found in the active layer, and we are informed that the materials in all layers must be well matched in terms of the atom-to-atom

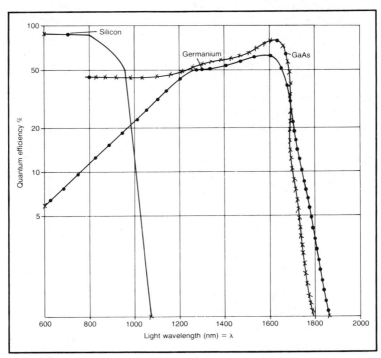

Fig. 1-5. Some relative quantum-efficiency curves for some types of PIN diodes (courtesy of Northern Telecom).

spacing in the crystal lattice. This insures few defects in the junction region and makes the unit more efficient and gives it a longer life span.

In Fig. 1-6 when electrons flow into the active region they are excited and when they return to their normal energy levels light is given off in the process. The wavelength of the light so produced can be governed by the material used in the solid-state device (especially the material used in the active layer). More energy is used from the electrons in producing short wavelengths (high frequencies) of light than for the longer wavelengths. Notice how the light is produced at right angles to the active surface in Fig. 1-6 for the LED and in the plane of the active material for the laser device.

SOME MODERN CONCEPTS AND
IDEAS ASSOCIATED WITH FIBEROPTICS

We find that if we are going to send intelligence over optical fibers—which are of a special type and specially made so that the light rays will pass "down" them (instead of out through their sides) that we must first have some fibers of the necessary purity and composition to pass this light easily and without much attenuation. That statement is almost too obvious to write, but we need to have it down in black and white. Second, we need some kind of a light source which will emit a bright and highly directional beam so we can get it into these optical fibers as efficiently as possible. Remember that the fibers are very small in diameter (a fraction of a millimeter in diameter in fact) and had to have a composition such that the light-losses would be less than 20 dB/km. Many current types have losses much less than this figure. The fibers have been made by Corning Glass Works by synthesizing silica glass. The raw materials were vaporized and deposited inside a length of quartz

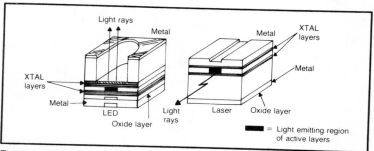

Fig. 1-6. LEDs and lasers for optical fiber applications (courtesy of Northern Telecom).

glass tubing which was then drawn into a small diameter rod. The modern concept of the method of conveying intelligence over a light-carrying transmission system is to use the Binary coding system or Digital transmission system. That means the source must be capable of being turned *off and on* very rapidly and the faster this can be done the more intelligence we can transmit. The receivers must be able to receive these "light pulses" accurately and develop electrical currents from them. In the pulsing process they must recover very rapidly from the "impact" of each light pulse. A sharp turn-on and turn-off of the receiving element is required to produce the pulse shapes needed for positive identification, and thus eliminate noise and any other spurious signals which might enter into the system.

You might well be wondering what about the use of ruby-rod type lasers or gas lasers as a source of light for fiber optical systems. Well, the truth of the matter is that the ruby-rod may be cumbersome and the gas laser may have a short life (it is also pretty cumbersome for this type application). So, the reason they aren't well thought of in this application is short life and size and weight etc. As we now know, the LED and solid-state laser as shown in Fig. 1-6 are most applicable to this means of intelligence transmission.

With the channel (the optical fiber) available in the required pure state and strength, and with the source light now available in LEDs and solid-state lasers, the requirements for the photodiode became apparent. We know, from our previous discussion that the *PIN* type diodes and the *APD* (avalanche) type diodes do the job. They have the necessary conversion efficiency (light to electricity) and they are small and have the quick recovery time required.

SOME SAFETY TIPS

If you plan experimentation with light producing sources which you may find applicable to optical fiber type systems—**be careful! Do not look directly into the light-emitting region of the unit. They will emit rays which are very harmful to your eyes and permanent eye-damage could result. So be careful—be safe—and be happy!** You can check the output of such devices with a safe and sane system which won't present any danger to you. Learn about this scientific method and use it!

HOW STRONG IS AN OPTICAL FIBER?

The theoretical strength of silica glass (tensile strength) is said to be as high as one million psi! Tensile breaks have occurred in laboratory tests at Northern Telecom at about 5 kilograms of force. Since 2.2 kilograms equals a pound, this means a pull of about 10 lbs (or slightly more). But, it is common at Northern Telecom to test such fibers to a level of 300 to 500 grams (one kilogram equals 1,000 grams) to weed out any weak fibers. Sometimes these appear and have cracks in them or other imperfections which would later cause problems if they were used as they are produced. So, fibers must be selected "selectively" after tests and measurements to insure that they have the correct properties—low-loss light transmission and the necessary tensile strength to stand up in installation and use. The testing used with the fibers has been very exacting: high temperatures, humidity, water immersion, and exposure to acids and other types of solutions they might encounter. It is now expected that their life will be equal to that of the well known copper cable lines now in use in telephone systems. The fibers are so flexible that you can wind them around one finger. During installation care is usually taken to avoid sharp bends and to avoid bending splices so as to minimize the stresses that might cause premature failures.

TRANSMISSION OF A MULTIPLE OF
CHANNELS OVER FIBER-OPTIC SYSTEMS

We have stated a slight bit of history in the transmission of multiple carriers, each modulated with sound or TV or other intelligence (when such information is conveyed by amplitude type of modulation as mathematically stated in Equation 1-4). It can be of value to us to examine a diagram which illustrates how such a multiplicity of channels might be transmitted over a fiber-optics system using the same kind of techniques used for carrier telephone systems. See Fig. 1-7.

We note, in Fig. 1-7B that the channels are actually well separated by "guard bands." Because of the carrier frequencies used, the addition of the sidebands does not widen the channel much from that required by the carrier alone. Thus many channels can be transmitted over the same wire pair or through the same waveguide (or the same optical fiber) and since we know that in optical fiber transmission the frequencies are very high there can be such a multitude of non-interfering channels. Practically all the telephone conversations on the North American continent could be transmitted over this kind of system at visual light frequencies without having one cause any overlap or interference with another channel.

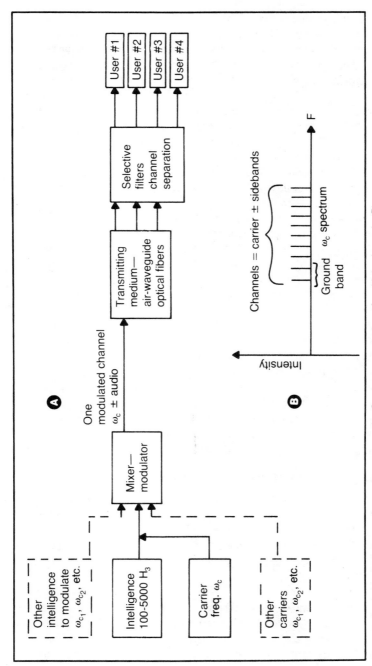

Fig. 1-7. Basic block diagram of a carrier system (A). Carrier transmission of multiple channels (B).

17

Of great importance in this system of intelligence transmission are the selective filters at the decoding end which separate the channels. Some thought has to be given to the various carriers so that intermodulation (mixing of the carrier frequencies to produce intelligence sideband frequencies which distort the original sidebands) does not occur. It is to be noted that there is a system using this type of technique which is called *suppressed carrier transmission*. Once the carrier has been modulated with the intelligence, the carrier itself conveys no information in an AM system and can be removed by appropriate filtering. This is as you already suspect the concept of single-sideband transmission much used in current amateur and commercial radio transmission applications. This would increase the number of bands of useful information which could be transmitted over a pair of wires, waveguide, or optical fiber because the spacing between channels then could be reduced. If the carrier is eliminated at the transmitting end, then it has to be reintroduced at the receiving end for demodulation as is done in current SSB systems.

The carrier system of transmission of intelligence has long been known and used. The generation of frequencies in the visible light region is a fairly new accomplishment and the production of sufficiently pure optical fibers with the methods of bending them, connecting them to input and output (I/O) devices, and splicing them is a reasonably new development. Like many things scientific, the principles may have been known or suspected for a long time but the physical implementation of those principles may have had to wait upon technological development of devices before such concepts could be put to practical use. Such is the basic story with optical fiber transmission of intelligence.

LET US EXAMINE AN OPTICAL FIBER MORE CLOSELY

What is so great about the optical fiber? It is a piece of glass which permits the passage of light. Right? Wrong! It is *not* just a piece of glass. It is a very fine strand of very special glass which may be only 125 microns in diameter! A micron is one-millionth of a meter. It is a glass strand that is about the same thickness as a human hair.

It has been proven that the electromagnetic waves that constitute light (and we know this from the way we can handle them and the way they behave) will tend to travel through a region that has a high refractive index. So, we make the center of the glass strand—the *core*—that type of material. Some glass fibers have a core diameter of only 50 microns and this core has a *graded* type of refractive index. The importance of this grading is to make the core a little more broadband than a core would be if it was the same index of refraction throughout.

Now that we have the core, in order to keep the light inside that core, we need to cover it with some kind of material with a different index of refraction. If we don't, the necessary *reflections* from the junction of the two materials won't take place. So, there is formed on the core another coating that is called the *cladding* and this has a lower index of refraction than has the core itself. Finally, to make it all look professional, make it stronger, and prevent damage to the cladding, it is common to form a "jacket" over the cladding which is usually some type of plastic material. The only purpose of the jacket is protection.

The story doesn't end here. We have to consider the transmission of digital light-pulses at very high rates through this fiber and we want to learn how multiple conversations, pictures, and so on might be sent over these optical fibers *simultaneously.* We want to know more about coupling light into such a fiber and what type of connecting devices are necessary to get the light out of the optical fibers.

WHY IS SNELL'S LAW IMPORTANT?

We've been discussing reflection and refraction. Willebrord Snell was the man who discovered the law concerning this phenomena in 1621. Imagine a name such as "Willie" had! Well, in any event, what he came up with is an equation which has lasted for the past several hundred years and this equation is important to our concept of how light rays travel down our "light tube." The equation or Law is commonly stated as:

$$\text{Snell's Law} = n_1 \sin (\theta) = n_2 \sin (\phi) \qquad \textbf{Equation 1-6}$$

which, in its simplest terms, simply says that the *ratio* of the refractive indices is equal to the *ratio* of the sines of the angle of incidence and the angle of refraction. Since this is somewhat difficult to visualize mentally, let us look at Fig. 1-8. We see that a light ray coming through the cladding into the fiber-optic core material is bent less than the angle of incidence. Or, to state it another way, the angle of incidence is greater than the angle of refraction.

We must look at Fig. 1-9 to see the significance of this concept.

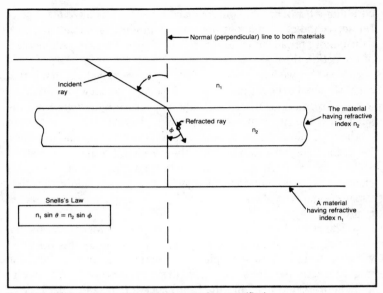

Fig. 1-8. Basic incident and refractive angles specified.

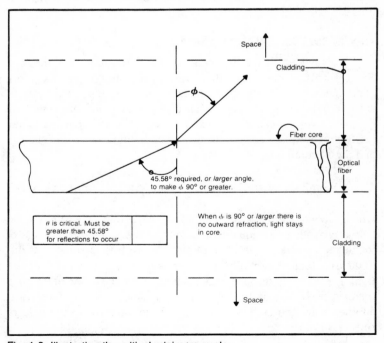

Fig. 1-9. Illustrating the critical minimum-angle.

We turn the whole thing around so that we now have a ray originating (somehow) inside the core material. It goes upward toward the cladding through an angle that we have shown to be 45.58 degrees (even if it does not measure this with your protractor). Assume it is this value. This is called the *Brewster angle,* or the *critical angle* and this is important because *if the angle is any less than this value* some light will go through the core into the cladding and escape from the "light tube." We don't want that to happen. Notice what happens when the angle becomes less than critical. We have a more direct ray going upward (as shown) and so, as you would suspect, instead of "grazing" the boundary between the two materials (the core and the cladding) the shaft of light passes through that boundary and goes from the core into the cladding area and thence may pass outward into space depending upon the angle at which that ray strikes the boundary between the cladding and space!

Be sure you note that the angles are measured *from the normal line to the ray.* Thus if (ϕ) has a value of 90 degrees it is *parallel* to the side of the core, or as we say, it remains inside the core for this value or a greater value.

DO DIFFERENT MATERIALS HAVE
DIFFERENT INDICES OF REFRACTION?

Different materials do have different indices of refraction. For example if a reference material is empty space and it is said to have an index of refraction of (1) then water has an index of refraction of 1.33, and crown glass has an index of refraction of from 1.41 to 1.61 depending upon its composition and manufacture. It has been found that a diamond has an index of refraction of about 2.4. One would suppose that the various facets or polished "flats" on a diamond are specifically angled to help produce the greatest outgoing reflection of light rays which are refracted internally from these "flats" unless they are already going outward in a desired direction. Thus we get various magnitudes of diamond "brilliance."

WHAT DO THE TERMS SINGLE MODE
AND MULTIMODE MEAN IN FIBEROPTICS?

A fiber is said to be single-mode when certain relationships hold. Actually, what is meant is that the fiber *operates* in a single mode, it does not *have* a mode based on its shape or material. The relationships are stated in equation form as:

$$\frac{2\pi a}{\lambda} \quad \sqrt{(2n)(\Delta n)} \leq 2.41 = \text{single mode} \qquad \textbf{Equation 1-7}$$

where: the core radius is (a)

the wavelength is (λ)

the core reference index is (n)

and (Δn) is the difference in refractive index between the core and the cladding.

The equation can be manipulated to find the permissible size of the core for a given size and type of cladding if the difference in the core and cladding refractive indices are known. What this tells us is that in the single mode only one frequency of light is considered to be transmitted.

An optical fiber is said to be *multimode* if either the core diameter or the cladding refractive index of refraction are larger than the limit as stated by the previous equation (1-7) for single mode operation. In multimode operation many different light rays (each traveling at a different angle of reflection but at less than the *critical angle*) will be guided down and along the core. In the equation stated, it is assumed that the index of refraction of the core and cladding are uniform and the change in the refractive index at the boundary of the two is abrupt. It is possible to have a *graded* material such that there is a gradual change in the index of refraction from the center outward. This, it is said, decreases the mode dispersion along multimode light channeling fibers.

LOSSES IN OPTICAL FIBER SYSTEMS

When we put light energy into an optical fiber (and do it in the proper manner so that it is transmitted by reflections down the core or along the core material) we find that inevitably the light energy weakens as it travels over any distance. This is due to the attenuation of the signal by the fiber material which absorbs some of the energy in the passage of the light rays. It becomes very important to consider just how much loss can be tolerated in the various types of systems which might use optical fibers for the transmission of intelligence.

We know that in some automobiles optical fibers are used to convey a small amount of light from the tail-lights or the headlights to small panels in the front or at the rear of the inside of the car. By glancing at these panels you can see whether your lights are on and functioning properly or not. There are other applications such as in airplanes, ships, computers, medical instruments, and in buildings wherein optical fibers might and are used for signaling, intelligence transmission, and even control applications. We consider the losses tolerable in these types of short-range applications. The studies show that losses of up to 50 to 100 decibels per kilometer may be permissible in these applications. In other applications such as in telephone systems where very long "runs" of optical fibers are imagined and used (distances of up to and greater than 14 kilometers) the efficiency of the fibers to transmit light rays must be better with losses equal to or less than 5 decibels per kilometer. Some types of fibers currently in use have losses of only 2 to 3 decibels per kilometer when the light frequency is around 800 to 900 nm (nanometer). This is called the "short wavelength" region by those industries engaged in developing this means of communications.

Consider that in long span arrangements or systems the light might be reinforced at various stations along the way. One would not want to have so much reinforcement (so many repeater stations) that the cost becomes prohibitive when optical fibers are used. As always, industry is interested in "cost effective" systems that will show a profit in some manner.

Now that we have brought up the subject of frequency and wavelength along with attenuation of the signals, we need to expand slightly on this idea by saying that some frequencies will be attenuated more than others in a given type of fiber. We find ourselves considering some of the many factors related to optical fibers, such as, its strength, reproducibility, index of refraction, losses, costs, repairability and so on. One important item is maintaining the transmitting frequency at the proper value to "match" the optical-fiber system being used. Some lasers are affected by temperatures but they can generate the required energy. Lasers are highly directional. LEDs can generate about as much power when properly excited and fed strong enough currents but have a slower response time than lasers. It is said that the complexity of the laser system is much greater than that of the LED light excitation system. Reliability of the source is also important. The "lifetime" of each type of device is becoming an area of study. It is said that lasers can have lifetimes of 10^5 hours and LEDs may exceed this by some 10^2 hours of operation. The losses in the optical fibers are from *Rayleigh scattering* at the shorter wavelengths and by infrared absorption at the longer wavelengths. Rayleigh scattering tends to diffuse the light so that its intensity is not that of a single point source and thus it is weakened.

SOME IMAGINED APPLICATIONS OF FIBER-OPTIC SYSTEMS

While we digest some of the technicalities of the past few pages, let us examine some of the possible and projected and imagined uses for optical fiber systems according to some very inventive and imaginative (and possibly sometimes incorrect) science writers. One science reporter imagines that it might be possible in the future to send an optical signal to your home from your car to cause the oven to turn on so dinner will be ready for you when you arrive. Or, perhaps you want to turn the house lights on or off. You may be able to do it from your car wherever you are. It is said that a *telecontrol* system which uses optical fibers is operating in the western part of Tokyo and being a part of the telephone-telegraph system reaches into homes and business and all such places where special signals might cause special machines or devices to operate to do something nice for you. What seems to be important is that the regular "wires" and cables of the telephone exchanges will one day be replaced by the infinitely smaller optical fiber lines. These can transmit even more information, control signals and pictures much easier and more accurately and with less problems due to interference. There will be a "fiber-optics revolution." This means an expanded operation from your home, car, or office to do many things that might currently require either your presence or communication by the slower and more uncertain means used at present. Imagine getting your bills before the ink is dry on the purchase request! But, don't worry—or "not to worry" as the saying goes, because at that same moment your bank will be extracting from your account a payment and depositing it in the seller's account. If anything makes you ill you could simply connect a bunch of gadgets to yourself and send all your symptoms to your doctor whose computer will tell you what to do about them!

Another report states that optical fibers are appearing in conjunction with new sensing devices. These devices range from gyroscopes that sense inertial references to heat sensing devices of various types and kinds even to the type which senses the temperature and operation of the motor of your car and then flashes on its LED-type display panel the delightful information that you'd better stop and do something about some problem!

We know that sensors have been developed to measure strain and stress in materials from airplane parts to buildings (but not in humans unfortunately) and it is believed that optical fibers can be used to convey information about the inner workings of nuclear reactors from the inside to the outside of such devices. We know such fibers convey light into and pictures out of humans, in medical applications and that the oil industry is using such devices with appropriate sensors to find that black "liquid gold!"

Because these fibers may be sensitive to changes in stress and strain, temperature, rotation, pressure, and perhaps even sound waves, they may become a part of the sensing systems of the future. All that has to be done is to know what form the light rays have when entering the fiber and then be able to determine what changes have taken place as these rays go through the fiber. Such changes are caused by the above mentioned areas of sensitivity. When one knows how the fiber changes with an increase or decrease in temperature, then one can easily (with a computer) calculate the changes to the light rays and thus have a means of measuring and sensing such temperature changes and evaluating them.

Sometimes it is necessary to split the light rays into two sections. This is true in sensing rotation. If we modify the speed of light transmission by adding the speed of rotation of an object (using a device that rotates in the direction the body is rotating such a device being a long fiber-optics tube), then when the original frequency of the light beam is compared with the half that was "split off" and sent through the fibers that are being rotated, there will be a slight difference in frequency which is proportional to the speed of rotation of the fiber container or housing. So, we can measure speed of rotation of something. This application is a new type of gyroscopic sensor and it can be applied to almost every type of moving object.

The ability of the fiber-optic transmission system to reject interference from electromagnetic sources can be very meaningful. That means greater reliability in control and in communications. In control applications (wherever they may be installed) slight static pick-up of electrical impulses can cause problems in the movement of machine arms, in the control of electricity, in power plants, and so on. This means of communication to and from and between electrical and electromechanical devices becomes of great importance in our future.

No one is certain just what the future of optical fibers is going to be but everyone is certain of one aspect of this new technology: it will cause a revolution in the electrical, electronics, communication, and computing fields. Its applications are growing at a rate which almost surpasses that of the solid-state integrated-circuit field.

In the field of robotics, optical fibers are being developed to give robots the ability to "see." Optical fiber scanners can be made up of a large number of individual fibers which might be only 0.003 inch in diameter (0.07 mm in diameter approximately) and these fibers are each fitted with appropriate I/O (input-output) connectors to get the light rays of a suitable frequency into and out of the fiber "light-tube."

It is said that these "bundles" of fibers can be arranged in a couple of different ways. One is to produce coherent rays of light by arranging each fiber so that it has the same physical orientation with respect to all other fibers in that bundle. This is the type arrangement commonly used in a fiberscope, which is flexible in construction. The opposite (the noncoherent type of fiber bundle) has fibers arranged in a random manner without regard for the orientation of the light rays or wavefronts emerging therefrom. Recall that if a laser is used as the light source, it can provide a coherent type of light, that we imagine to be a light ray with all wavefronts parallel and in a given orientation. If the fiber is now twisted, that wavefront will twist with the fiber. It maintains a constant orientation with respect to its source and the fiber material through which it passes. Noncoherent light is random in dispersion.

It is said that fiber optical systems can provide very precise control of a robot's arm movements, hand movements, etc., since they measure so precisely the position or movement of these extremities. Also, the light rays used with optical fibers provide the simple light-dark or one-zero logic which is so easily adapted to computerized control as a natural way of sending and turning the light on and off at the source. Of course, one could imagine an *interrupted* light beam as having meaning also (recall the photoelectric and infrared announcing systems such as sold by Radio Shack and other electronics suppliers).

It is imagined that because they are so small and one might use so many of them, that optical fibers might give some idea as to hand pressure exerted by a robot, or shape of objects as "seen" from the hand position. See our books on robotics TAB book numbers 1071 and 1421 for a more complete discussion of robots and this phenomena. Fiber optical *switches* are easy to imagine, you simply move the light-carrying fiber a little away from another so that the light does not go through the second fiber. That is an *off*. Moving the first fiber back so it sends light into the second fiber then makes the light path continuous to the detector and this becomes (or can become) an *on* position. Of course the *off* and *on* positions can be changed by reversing the wiring.

Because they are flexible, optical fibers may be used with moving parts of the robot without problems and one might use two such optical fibers to guide a machine tooling device along some kind of pattern or "edge" of a work-piece. They can give the robot "long range vision" as the actual "pixel element" formation unit doesn't have to be at the work end of the fiber which is scanning the illuminated object.

It is said that one of the advantages of using optical fibers to transmit light rays to "dark" areas and thus illuminate them (so say a doctor can look around and see what is going on inside of you—or a scientist can look inside some device and see what is there) is that "cold" light can be used and used with high intensity. If a large area is to be illuminated then many optical fibers would be used and placed and adjusted so that shadows and dark areas would not exist. Heat is generated when the "conversion device" (the light bulb) is near the work object because not all of the energy put into such an element results in light rays. Some of that energy is converted into heat. When you can have the source of the light a long distance away from the object, the light rays as conducted by the optical fibers will not have that heat content (unless they are in that region of the spectrum where heat is automatically produced with the ray) and thus we have what is called "cold" light. Fluorescent light tubes produce "cold" light but there is a lot of heat generated in the transformer which supplies the high voltage that is used with such devices.

Since the small size of the receiver and transmitter unit of a light system permits close proximity in confined spaces, the various applications of light are multiplied enormously. In Fig. 1-10 we see how such units might be combined in a single block to measure some size, shape, or movement of some machine such as a robot. One might decide after viewing Fig. 1-10 that the robotic hand has optical fiber nerves that connect to the LEDs and the photodiodes (receiving units) and these nerves then connect to various amplifiers, computing elements, and integrated circuits that can analyze and determine what the robot's fingers see, how they are grasping something, where that something is with respect to the robot's hand, and so on. Actually, each LED shown might be a dual light source which then gives a differential output for each finger. The possibilities for hand instrumentation using fiberoptics and associated devices is almost unlimited.

It is interesting that many professionals feel that the use of LEDs, lasers, and infrared beams of light that are conveyed by optical fibers may be the answer to some perplexing problems in connection with the transmission of light. Infrared beams are not subject to interference by ambient light rays. They work equally

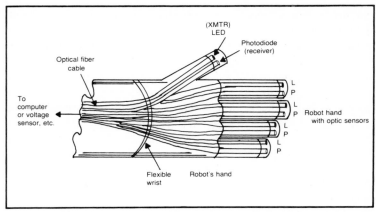

Fig. 1-10. A block of transmitter/receiver light elements in an application in a robot's hand.

well in daylight and darkness. They are invisible to the human eye. Thus, if the light changes in an environment wherein the sensor is used, that makes no difference in operation of the machine. Some believe that small amounts of dust may not affect the transmission of the infrared beams. If the part being handled is hot enough to emit infrared radiations then it is easy to follow and track its position with infrared sensitive sensors.

THE DISPERSION PROBLEM WITH OPTICAL FIBERS

We have indicated and stated that there are two possible modes of transmission of light in an optical fiber. One is the single mode. The second is the multimode. One equation that might indicate how many modes might be propagated in a given optical fiber can be stated as:

$$f = \frac{\left[\dfrac{\pi d}{\lambda} \sqrt{(n_1)^2 - (n_2)^2} \right]^2}{2} \qquad \textbf{Equation 1-8}$$

n_1 = index of refraction core
n_2 = index of refraction cladding
d = core diameter (meters)
λ = wavelength (meters)
f = modes propagated

Of course, as we stated earlier, if Equation 1-7 shows that only a single mode is transmitted, then this equation would not be applicable. Single mode propagation occurs with a very small size core (only a fraction of the size of a human hair). This makes joining a number of fibers together end-to-end (to make up a long "run" of such fibers) quite a formidable task. The dispersion problem, simply stated, is that various frequencies, and thus, wavelengths will be propagated through the fibers when conditions are not perfect. As you know, things can *never* be perfect. The reason many modes might be transmitted is because the source of the light, say a LED may actually create several or many frequencies within the very narrow band found at these tremendously high ranges. If the *guide* is so designed that it can sustain these other frequencies, and they are strong enough, then they will propagate through the optical fiber guide just as radar waves will propagate through a length of radar waveguide, if their frequency is above the cut-off value.

Mode dispersion problems can affect the transmission of any intelligence transmitted by means of "light pulses." This is sometimes referred to as the *"impulse response"* capability of the optical fibers. Mode dispersion and material dispersion tend to broaden the light pulses time-wise and even though you might have begun the transmission of intelligence with short and well spaced pulses this type of dispersion can cause the pulses to occupy a long time interval and thus reduce the time space between them until in the worst case the pulses actually overlap and there are no pulses and spaces existent at all! There is just a solid continuous ray of light at the exit. *Dispersion* then relates to the speed of propagation of the various frequencies within the transmitted band of frequencies in the optical light guide.

We find much literature which talks about "modes." However, the precise definition of a "mode" is lacking. It is true that a mathematical equation might be used to describe a "single mode" or a "multi-mode" optical fiber system, and after analyzing such an equation we find that this means that one or several frequencies or wavelengths seem to be propagated depending on whether the system is "single mode" or "multi-mode." A dictionary definition of "mode" is: "A particular form or variety of something." It might also mean "style." *Mode* and *dispersion* are important terms in fiber-optics so let us have another "go" at understanding them.

We have to think that "mode" is related to the number and kind (length) of wavelengths that might be propagated through the core of an optical fiber and that the mode is related to the velocity of propagation of these wavelengths through that core, normalized, as they say, or related to a one kilometer length of fiber-optic material.

If we know that a LED used to excite a beam of light in an

optical fiber actually *sends a band of light frequencies* into that fiber, then we have no trouble grasping the idea that there are various wavelengths or "wave fronts" moving inside that transparent material. We should also have no problem considering the fact that some of these waves may travel in a direct line through the fiber and some will "bounce around" by being reflected from the junction of the core and cladding. These will require a longer time to get from the input end of the fiber to the output end of the fiber. It has been shown that a LED may have a band of plus or minus 25 nm about a center frequency. It generates this band just because of the way it works. So we *do* have a number of wavelengths trying to go through the optical fiber when it is excited by a source device. Some do not get through because they enter at such an angle that the *critical angle* for propagation is exceeded, and so they escape into the cladding—and, like Clementine, are lost and gone forever! But lots of others do make it into the fiber, entering inside the required *acceptance cone angle,* and so do propagate in one way or another along the fiber length. Becoming very technical for a sentence: *Modes are discrete incidence angles* for the light rays. A glance at Fig. 1-11 explains this further. In Fig. 1-11 we find a direct ray going straight ahead from input to output (a). Then we find some captured rays which hit the junction between the core and the cladding walls at less than the critical angle for escape and so are reflected (b). Some rays are too direct and having passed the critical capture angle escape into the cladding "and are lost forever" (c). It isn't hard to look at this sketch and say, "if there was just one light frequency (such as from a laser) then it might be the direct ray, and since it passes down the core, it is single mode." But, perhaps this is optimistic. Is there *anything* which will produce just one frequency with no harmonics? On the initial ray from the course (c) we have taken the liberty of drawing it as a direct-line from the source. Actually since there is an index of refraction between the air and the core, there is a one-bend "kink" in the first part of ray (c) which is not shown. Bear with us if you are a perfectionist. Equation 1-8 stated earlier relates some important optical fiber quantities to the number of modes which would be propagated in a given fiber size, etc.

We find that since the higher order modes travel farther (due to reflection or bouncing around) they must arrive at the output *after* the lower order modes arrive. In this context a "higher order mode" must be a lower frequency since it has a longer wavelength and thus bounces around more—or could it be that it is a higher frequency

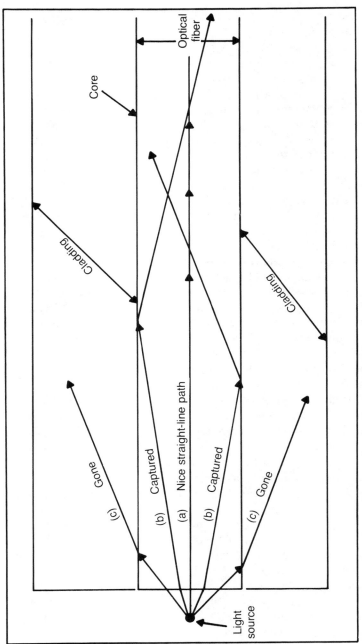

Fig. 1-11. The transmission modes are various incidence angles of propagation.

31

which produces a shorter wavelength and this is refracted more? Think about it!

In any event, we recall from radar history and technology that when one desires nice sharply-shaped pulses one needs many frequencies all put together. In this context, if we say that some frequencies from a pulsing light source will not get to the fiber output at the same time that other frequencies do then *we have some delays* in transmission. That has to affect the pulse shape at the output (make it rounded, less well-defined and so on). This is no problem *if* we are using a slow transmission of pulses, but, if we are in the 50 megabit range or higher then we've got a problem. Delays such as this cause errors and loss of pulses and pulse shapes and all kinds of other bad things.

Are you ready for this? Delays like this have a name. These delays are called *chromatic dispersion*. Simply stated this means: variable delays due to various wavelengths trying to propagate through the optical fiber. It has been found (Northern Telecom) that pulses of light in the 840 micrometer region will be broadened about 0.1 nanosecond (10^{-9}) per kilometer of fiber per nanometer of source width. Thus, they say, that a 5 nanosecond pulse from a LED which has a 40 micrometer spectral width will, after going through some 5 kilometers of optical fiber, broaden out to about 20 nanoseconds. A laser source pulse which has a 2 to 4 micrometer width will gain a broadening effect of about ½ microseconds. Now we have a basic idea of at least one problem in the transmission of high speed pulses with light—the delays caused by the fiber material and the different transmission times of the various independent frequencies through the fiber, resulting in changes such as broadening of the pulses and compromising the pulse shapes in this method of transmission.

There is another effect called *multimode dispersion*. This is a concept which we can grasp quite easily if we remember that there is such a thing as interference between one wave and another in the electromagnetic spectrum—which, of course, encompasses the visible and invisible light frequency range. We can refresh our memories on this kind of happening if we remember that radio waves are reflected from the Heaviside layer (if the angle of incidence is less than the critical angle) and they are reflected and fall upon various places on our planet and they add to or subtract from other waves that also arrive at that particular destination. When we have an add-to situation, we have an increasing intensity of signal and in our listening device the sound becomes louder (usually). When we have a subtract-from condition then we have what we call "fading" and if conditions are exactly right, then we can have a complete elimination of the signal for a short or medium period of time at that location.

In an optical fiber we have that same condition existing but at the frequency of light. The same frequency waves tend to add to and subtract from each other as they progress down the fiber's length and so when we consider what remains we find that only a given number of "paths" or "modes" exist in the fiber which permit transmission completely from input to output of the fiber. Each path for each different frequency is different and so each path (direct or reflecting from the walls of the core) will have a different time of arrival at the output—or a delay. Then we have the same bad conditions caused by *chromatic dispersion.* It was found, (by Northern Telecom) that when the core material was of uniform index of refraction this kind of delay could amount to as much as 20 nanoseconds per kilometer of travel. To state it much more scientifically, it amounted to about 20 MHz/kilometer roll-off of frequency meaning about a 3 dB optical bandwidth. This in turn, as we have stated, meant the loss of those wider bands of frequencies that are necessary to make up a precise pulse shape. Recall that a "square-wave pulse" mathematically requires an infinite number of frequencies, harmonics, etc. to be a "square-wave pulse." That says, in turn, that we need an infinite bandwidth. So, if the bandwidth is reduced, the pulse shape is affected. Recall that in radar it is said that the bandwidth required is

$$\frac{1}{\text{pulse width}}$$

and this little expression gives us some idea of what is going on here in the narrowing effect of the multimode dispersion in optical fibers. There is a word used in connection with this effect and the word is intramodal. It is found that if a graduated system of "index of refraction" is fashioned within the core material, this may result in increased bandwidths as high as 3 to 6 GHz/km with some experimental types of fibers.

We get some good and some bad effects in the transmission of light rays through an optical fiber core. We have mentioned some of the bad effects, now let us consider some good effects. There is a "happening" called *mode mixing* which can take place across splices and within the fibers themselves. You recall that we have stated that when the paths of the various light frequencies are different it takes different times to get different rays from the input to the output. In "mode mixing" there is some interplay between the various modes (paths) in such a manner that the time it takes all rays to get from input to output tends toward an "average" value. That means that the delays tend to balance out and all rays tend to arrive at the output at the same time. How nice! Well, to be honest, you cannot have everything, and so there will still be *some* delays, but these won't be as bad as they might be if we didn't have this "mixing phenomena."

What is nice, is that you can induce mode-mixing by "messing-up" the light-fiber path. Using microbend effects, which result when you twist and bend the light fibers really confuses the rays so they don't know which path they had a moment ago and tend to use new ones and different ones so that "mixing" occurs. The problem is that increasing the path length this way and *mixing-the-modes* (nice sound to that phrase isn't it?) causes some losses. Well, it is bound to happen, when you confuse the path and some rays don't know where they are going, they just don't go or go incorrectly and get lost—what'd you expect! So, it is a "trade-off." You have to consider the "nice effects" along with the attenuation and the sources of light and so on and then decide if you want to improve this effect or not. Perhaps more study and experimentation will determine the proper way to go. You should remember that *intramodal* is a term meaning inside a fiber mode or along a fiber ray path, while *intermodal* means the various paths from one fiber to another.

In most college physics courses there is much time devoted to light and the use of diffraction gratings and lenses and the concepts and mathematics and experiments concerned with optics which are appropriate to our discussion. We recall that one such series of experiments involves a prism and diffraction grating (involving Young's experiment) to determine the *lines* of the optical spectrum. We now think of each line as a band of light frequencies, so closely related that they seem to merge together to become one central frequency or line. It is common to describe the light rays transmitted through an optical fiber as being in "lines," meaning the wavelengths which actually pass from input to output through this type medium. We again refer you to any good, standard college textbook and highly recommend a serious study of Young's experiments if you are looking forward to a serious study of optical fibers. *Michelson's* interferometer also is recommended as a basic study in that our modern laser gyroscopes are based in principle on this experiment.

But, back to the use of a diffraction grating and/or a prism, when an incident light is focused on such devices, they will "spread out" the light spectrum such that one can see the various bands, or lines of light. The prism can produce colors which identify with various frequencies. The diffraction grating will produce light and dark lines showing the reinforcement of rays or cancellation of rays as a function of distance. The resulting patterns can be analyzed to obtain various information about the light ray being propagated. Diffraction gratings are often used to measure wavelengths and to study the structure and intensity of spectrum lines. This can be important in the study and experimentation efforts concerned with optical fibers.

It is important to mention that with the ultra-high frequency radar systems now in use in various applications, *dielectric rods* are often used as waveguides and antennas. The principles are the same as for the region of optical wavelengths and we should then turn to a study of the use of these devices and the theory thereof in radar applications.

With optical fibers, even of a transparent plastic, light can be "piped" from one location to another by introducing the light into the fiber at one end in such a manner that total reflection will occur at the output boundary of the rod (where the cladding or strengthening material joins the rod surface) *if and only if* the light is introduced at an angle below that of *critical.* We have noted that the critical angle—that angle at which the the light "hits" the boundary edge of the two materials (be they plastic and air or anything else) is *less* than that which will permit the light to pass through that boundary. Mathematically that critical angle in terms of the material is:

$$\text{Sine (Critical Angle)} = \frac{n_2}{n_1} \qquad \textbf{Equation 1-9}$$

$n_2 =$ second medium index of refraction for air equals 1.00
$n_1 =$ first medium index of refraction for glass of a given type is 1.50

Thus the critical angle is that angle whose sin is $\dfrac{1.00}{1.50}$

or 0.667 and the angle is 41.7 degrees. This means that if the light rays hit the boundary or edge of the glass rod at an angle, measured from the perpendicular across the rod, *at less* than this angle, the light will escape through the rod's edge into the second medium. See Fig. 1-12.

Unless the *incident angle* is 90 degrees (at which angle all the light rays pass through the rod into the second medium and escape *and do not* travel down the rod) some light rays will be reflected

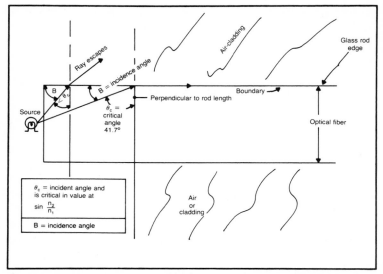

Fig. 1-12. The critical angle of reflection inside a fiber-optic waveguide.

down the rod even if the angle is less than the critical angle. But the closer the incident angle gets to 90 degrees the fewer the number of rays which will be reflected, until at 90 degrees none go down the rod length.

Make a note that since *images* are variations in light intensity which impinge on our eyes, if we can cause these variations in light to come back through an optical fiber or a bundle of optical fibers then we can "see" things which the other end of the fiber is pointed at, providing that the objects there are properly illuminated. This is the basis for the medical instruments which permit examination of our body's interior. One bundle of fibers may route the light rays and another bundle of optical fibers can be used with magnifying lenses to see what is going on inside.

When we consider the transmission of images using optical fiber bundles we have to remember that normally each fiber will transmit only a small segment of the image. Thus it has been common to use these "bundles" of fibers (in which each one carries only a small part of the image) to convey a scene from one end of the fiber length to the other end. The ends of the fibers on the receiving end are all viewed simultaneously and so the image can be seen. This is like building up a picture on your TV screen by using a multitude of dots. We aren't aware of the dots. When we view the screen we simply see the total image.

Since the fibers may be flexible, and since they are very small in diameter, they may be clad in some strengthening material so stresses and strains won't affect them. If you've visited any gift shops in recent years, you have probably seen a display of an "optical-fiber-tree." This unit, rather pretty in concept, with its flowing lines of almost clear transparent material emerging from a vase or jar or whatever, has at the tip of each fiber a glowing dot of light. That is the reflection of the light source inside the container, and it appears *only* at the end of the fiber. As we inspect such an ornament we would then get some idea as to why so many fibers are necessary in a "bundle" to convey a total image from one end to the other. Next time you're browsing gift shops and curio shops, look for such a display and examine the fibers as to size, material, light emission and so on. You'll find that there is a "light loss" and this can be due to the absorption of light rays by the material itself.

Considering the image transmission problem again, be aware that it has been found that if some fibers are "tapered" so that they have a large-end and taper down to a smaller end, the larger end can be placed to view the image and as the light rays come back through the fiber the image is carried intact and can be seen with relative ease using magnifying lenses that may or may not be attached to the viewing end of the fibers physically. There is another phenomena which occurs in the transmission refraction and reflection of light rays. This is called "double refraction." It is important because it can mean that a single ray may be split and have two sections or beams that are polarized at right angles to one another. Study any good college physics text for details of this phenomena. Here, we want to point out that the fact that light can be polarized can be of utmost importance. This means that we can design "light tubes" that will accept only certain polarizations and not others, thus we get "light selectivity." Also, if we polarize two beams from the same source so that they have equal and opposite polarizations, then cancellation can take place when they are combined. This effect may be useful in a light communications system. A *calcite crystal* can polarize two emerging rays from a single ray so they have a 90 degree polarization with respect to one-another. An optical fiber system must not cause cancellation of the rays within itself or the losses would be too great to make such a transmission system practical. We shall examine some communications applications using optical fibers in the next chapter.

Chapter 2
The Use of Optical Fibers in Industrial Applications

One of the most important considerations in the use of optical fibers and light rays for communication and data transmission is the fact that light rays are almost immune to electrical interference when sent over an optical fiber transmission path. Electromagnetic radiations such as sparks, lightning, and crosstalk effects are practically eliminated as interference sources in an optical fiber transmission system. Think how important this is! If we have a large computer installation, for example, and we want communication among and from the various machines, and the integrity of the communication is dependent upon elimination of interference signals from the system. Using light and optical fibers as the medium solves this problem.

We must consider the meaning of the use of light and light frequencies in communications. If we consider the fact that a small band of frequencies (kilohertz width perhaps) is needed for transmission of intelligence, then just think of how many such bands can be contained in the light region of the frequency spectrum without interference from each other. Also, since the bands might be made wider, information can be transmitted at much faster rates. Gigahertz rates and perhaps even higher rates might be used and still we have sufficient bandwidths to handle a very large number of simultaneous channels. It has been found that in optical fiber systems one can send analog and digital data "side-by-side" without causing any problems. What this means to telephone companies is evident. Costs are less than with copper wires, less crosstalk and interference, and fewer cables mean that it is almost inevitable that sooner or later all telephone communications channels will use this means of transmitting data, telephone, telegraph, and video signals.

THE THREE BASIC ELEMENTS IN A FIBER-OPTIC SYSTEM

No one disagrees with the premise that there are three primary elements required in a fiber-optic system for communications. The first element is *the transmitter*. That is the unit which must generate the light rays and be capable of being switched on and off quickly and completely and/or modulated in some manner with some signals representing intelligence. The second item is the *optical fiber* which must have a purity and cladding and packaging so that it is strong and transparent to the light frequencies being used. It must be capable of being spliced and repaired when necessary (at a reasonable cost) and be able to convey the light rays a reasonable distance before a "repeater station" has to reamplify the light beams to enable them to traverse the total distance over which they must travel. In some cases, many "repeater stations" might have to be used. Third, the *receiver element* must reconvert those light rays back into analog or digital currents and voltages so that the user station can separate and use the intelligence signals being transmitted.

It has been noted that in some cases where control signals of a dc or low frequency ac voltage may have to be used in conjunction with the light ray information, such signals may have to be sent over copper-type lines. Light rays cannot transmit dc signals as such. However, it is entirely possible to send digital data in the form of very high frequency pulses over the light beams and then reconvert these into appropriate on and off signals using a decoding device at the receiving end. Slight variations in pulse coding, width, spacing, and so on might be used with light beams to establish control functions and directional signals for control of machines or devices at some remote location. There are now available on the market transistor integrated circuits which convert analog-to-digital and digital-to-analog signals. So, once we reconvert the light rays or modulated light ray signals into voltages and currents, the solid-state technology is here and available for our use to do whatever we want to do with the signals in this latter form.

OPTICAL FIBERS AS SENSING DEVICES

It is important that we recall that if the optical fiber is stretched, or bent (microbends), or subjected to temperature influences, or even subjected to the influence of electromagnetic waves which can change the *polarization* of the light rays in a manner proportional to the magnitude of the disturbance, then we have (in effect) sensors of those disturbances. If we can cause the effect on the optical fiber to be related in some precise mathematical manner so we can predict and understand the direction and/or magnitude of the effect, we can produce a quantitative output of light proportional to the magnitude and/or direction of the effect. For example, let us suppose that we place an optical fiber near a furnace or heat source such that as the heat increases, the fiber diameter increases, and thus changes the path length of the rays going through it and also changes the angle of reflection and refraction so that a change in the *amount of light* (or polarization of light) passing through the fiber results. Can we not then assume that we can measure the light intensity at the receiving end, and relate this in some proportional manner to the temperature impinging upon the fiber? Of course we can. So, we have a heat sensor.

In like manner we can *split the beam* of light passing through an optical fiber (such beam being produced by a laser so that the light is a beam with uniformly directed rays and with a given polarity— direction of the "$\hat{E}$" vector) and then use one half of the beam as a *reference* or standard to which no changes occur, and send the other half through some device which is influenced by the "outside world" phenomena such as sound, light, pressures, temperatures, motion and so on. Then by comparing the change in polarization of the affected half of the beam to the reference beam we can measure the difference in polarization change and relate this mathematically to the "world phenomena" changes. We *sense* the "outside world" phenomena in this manner.

Thus we find that optical fibers can and are being used as sensors in a variety of ways, for object positioning, determining object size and shape, to determine rotation direction and rotation rates, to measure temperature changes, to monitor and detect sound by either an interferometer or pressure-change method, and even to detect electric currents, and/or voltages. These latter sensors use the change in polarization of a split beam caused by the electromagnetic field of an unknown current and then comparison with the standard as just described to determine the magnitude of the current.

SOME DEVICES AND EFFECTS USED WITH OPTICAL FIBERS

There is an effect associated with light rays or beams transmitted through optical fibers that we should be familiar with. This is called *birefringence*. What the word means is that the fiber will transmit a given polarization of light rays of a given frequency faster than another polarization of those same light rays. Thus, when some effect happens to a split beam light which causes the polarization change in the variable polarization half of the beam, a phase shift, which can be measured quite accurately, comes about when the two rays or beams are again combined.

Electrical magnetism is said to be measurable through this effect. If a *magnetostrictive* core material is used, with an optical fiber wound around it, such that when the core expands and contracts or vibrates due to exposure to some magnetic field of varying intensity then the fiber changes shape slightly. This can cause some parts of the light beam passing through it to have a slightly different polarization than the rest of the beam. Recall that light rays are electromagnetic in nature and so the fact that they can be influenced by other magnetic and electric fields is not surprising. If through a proper *Wallaston prism* the parts of the beam with a given polarization can be separated and then compared, you have a sensing of the magnetic field intensity, frequency, and so on.

It should not come as a surprise to us to know that light rays can be polarized and that this polarization can be changed with the proper elements—be they crystals, lenses, prisms or whatever. We already know that in modern radar systems (see TAB book number 1155) ferrite magnetic cores are used to change the polarization of electromagnetic waves at microwave frequencies and many fine devices have been produced that work because this phenomena exists. Light rays, being an extension of the microwave frequency range, can be expected to behave in a similar manner although the control and polarizing elements may differ in form and shape and size and appearance and composition. Also note that phase and amplitude and polarization comparisons with microwaves are an accomplished fact. Extending this technique to light rays and laser beams is something you'd expect to happen.

POLARIZATION

Now that we have introduced the subject of *polarization* of rays of light (or to state it more broadly, polarization of electromagnetic waves of any frequency or wavelength) we need to examine this subject, perhaps, in slightly more detail. Light is said to be an electromagnetic radiation and is a *transverse wave* that has its electric and magnetic vectors at right angles to each other and to the direction of motion of the wave's propagation. Normally such a wave is described by the *poynting vector* shown in Fig. 2-1. The diagram shows the relationship between the electric field and the magnetic field and the direction of propagation of the wave front. It is to be noted that often the symbol ($\hat{B}$) is used instead of ($\hat{H}$) for the magnetic component of the wave.

It is the direction of the ($\hat{E}$) vector at any time (t) which tells us what the polarization of the wave is said to be. When all the ($\hat{E}$) vectors of a wave align in a given direction, or in a plane, that plane is said to be the *plane of propagation* of the wave, and in a "plane-polarized" wave, all such planes which contain the ($\hat{E}$) vectors of the light beam are parallel. Notice that the electric and magnetic vectors of Fig. 2-1 are extended at right angles to the direction of propagation of the wave. Even when these vectors vary in magnitude, as they do with modulation of any type or kind, they still have this orientation to the ($\hat{P}$) vector at any instant and all instants of time.

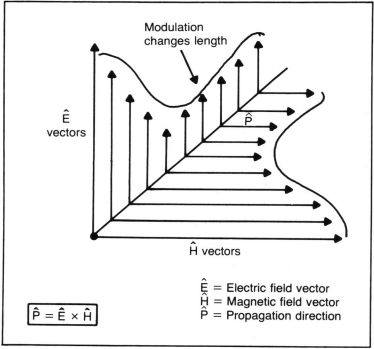

Fig. 2-1. A concept of the poynting vector of electromagnetic radiation with modulation of the (E) vector.

It is important for us to understand that under some conditions and circumstances, we can have a *circular polarization* condition that may exist *if* the plane of polarization of the wave is caused to rotate for some reason or another, around the $(\hat{P})$ vector. We can also have conditions where a single light beam may be split into two sections, each of which may then have a different polarization due to some external cause such as the sensing of some external and changing phenomena as described earlier. We then need to be able to determine the polarization and change of polarization of light rays in some manner.

In 1817 a scientist named Thomas Young, who was investigating the properties of light and sound (which produces a longitudinal, or pressure type wave) performed an experiment by which light rays could be investigated. He found that light rays had the *transverse* condition of propagation just described. What happened was that two of his contemporaries, *Arago* and *Fresnel*, were able to split a light beam into two sections or paths by allowing it to pass through a calcite crystal. This production of a "double beam" from a single beam at the crystals interface is called *double refraction*, or in simpler words, *double bending*. The two beams will have their $(\hat{E})$ vectors at right angles to one another, and so they will not cancel when combined. If the two $(\hat{E})$ vectors (one in each beam) were to be exactly equal and exactly opposite in polarity (180 degrees) they should cancel when combined or more precisely be recombined into one ray of light, when then wouldn't exist. The reason calcite will "split" such a beam of light into two rays is said to be caused by the fact that one ray inside the crystal will travel at a different speed and thus have a different refraction inside and entering the crystal than the other half, which tends to go straight through the crystal without much bending (see birefringence).

Thus, what we are finding out by this memory-refreshment in college physics, is that there are some crystals which can split and bend rays of light in various ways that we now find helps us to use light in its new roles through its fiber-optics channels. Of interest also is that various frequencies of light can be separated (colors separated) through the use of crystals and prisms of various kinds.

SOME DEFINITIONS

The terms *single mode* and *multimode* have meaning for us with respect to the transmission of light rays through optical fibers. It has been pointed out that if the optical fiber is very small in diameter (a few millionths of a meter or so) that under certain conditions that can involve the type of core material and the cladding material, the light rays will follow practically the same path through the core from one end to the other. This is called the single mode of transmission. In the single mode of transmission the integrity of the input polarization is not necessarily maintained but it *is* possible to have the polarization of the input remain constant during the transmission through the fiber (if the fiber is stressed properly when it is manufactured). That is, the fiber core is manufactured so that it will cause no major change in the light's polarization during transmission through that core. As we have stated previously, a *multimode* optical fiber has a larger core and the light rays will travel many different paths from input to output depending on their frequency and wavelength and insertion angles. We note that in much of the literature there are references to *step index* types of fibers and also to *graded index* types of fibers. If we think about this for a moment the meaning becomes clear. A "step index" means an abrupt change in the refractive index of core and cladding. A "graded index" means a gradual change in the core index of refraction that is accomplished by changing the core material in some graduated manner from the center outward to the cladding boundary.

DETECTING SMALL CHANGES IN THE PHASE OF LIGHT RAYS

There are many industrial and military applications of optical fiber systems that require very small changes in phase (that can exist between two light beams of the "split-beam" type) be detectable and measurable. This is no small task in itself. There can be phase changes due to reflection, there can be phase changes due to the change in the fiber structure through which the rays pass, and these latter changes can be due to such phenomena as water pressure in advance acoustic systems and magnetostrictive effects on substances that have physical contact and relationship with the optical fibers (heat and so on). There can be phase changes due to a change in the velocity term of the basic equation (such change as due to actual physical motion of the base structure of the optical fiber system as is found in the laser-type gyroscopes).

To determine these very small phase changes one uses an *interferometer*—and that brings to mind the famous *Michelson-Morley* experiments commonly discussed in most good physics texts. With light, one uses such a method and compares the phase of a variable-phase beam with the reference beam of a two-beam system. In this system of comparison one requires that the light beams be "coherent" or have the same polarization and this, as you'd expect, might be obtained from a laser. However, some studies point out that even then the coherence of the beams may change depending on the reflections (paths) internal to the optical fibers for the frequency being propagated.

Basically, the interferometer effect can be said to be a reinforcement or cancellation of the wave fronts. Therefore one could propagate almost any type of wave, split it, cause some shift of the phase of one beam due to something external or something to be measured, recombine the two beams and then see how much *interference* one beam has with the other under these conditions. By some analysis it is possible to relate the interference effect with numbers that represent the changes to the external world being measured so that you can quantitatively measure those changes.

The optical fiber system of light transmission is flexible to some extent. Also, if we put our minds to it, we can imagine that we could have a "splice" between two sections of optical fibers, such that when the two ends are perfectly aligned, the maximum amount of light is passed from one section to another. If anything causes any shift in this alignment, then less than maximum light transmission will take place. Aha! Now we have an idea for measurement of anything from physical vibrations to detection of objects in a robot's gripper! If one section of optical fiber is permitted to vibrate with the "outside force", the changes in light will be proportional in frequency and amplitude to that vibration regardless of its cause or source. The sudden darkening of light transmission from such sensors in a robot's gripper could mean to a computing section that the "gripper" has encompassed a part to be moved, located a part or whatever. The requirement here is that the light through one section of the optical fiber must be constant and the second section of the fiber must be permitted to vibrate with the external energy force in a small displacement (or be capable of transmitting the light rays through such an air path as would be necessary going from a robot's finger-tip to its thumb so that interruption of the beam of light can have the meaning that an object is intervening between those two robotic elements). The reflection of light rays (as in a grocery store "bar-counter") also has meaning in this context.

It has been determined that almost anything can be sensed using the phenomena of polarized light shift, or intensity shifting of the light rays, if the proper conversion element can be fabricated. One example of this is the so called *"Pockels effect"*, which is simply the rotation of the polarization of the light beam as it passes through a crystal which is subject at the same time to the pressure of an electric field due to an applied voltage across that crystal. You know how a piezo-electric crystal will "squirm" under an applied voltage. It is opaque to light so it won't work in this type application but there are some crystals of other substances which will be clear to light transmission and will cause some effect on the light rays due to the application of voltage to the crystals. *Lithium Niobate crystals* are one type that fill the need for this type of sensing device. They cause a rotation of the polarization of the light ray that is proportional to the magnitude of the applied electrical voltage. Now, all that remains is to use some kind of interference-effect detector system to compare this rotation amount to a standard that has not been rotated. Knowing the mathematical relationship between the rotation of polarization in degrees and the voltage causing it, if we can measure the amount of rotation it is a simple task to determine what voltage was present across the crystal. We can by using interferometric methods determine that polarization rotation.

FIBEROPTICS IN CARS AND AIRPLANES

One would expect to find that the automobile manufacturer's have found a use for optical fibers. They use these "light guides" to convey information from a headlight or taillight to a dash or "above the rear window" indicator (and end view of the optical fiber is the indicator) to let you know that that light is operating when turned on. Also because optical fibers can be adapted to dashboard illuminating panels, they become attractive there in that capacity.

Modern aircraft must have instrument readings that are as immune to outside or transient interference as possible. Optical fibers as the channel along which such information flows may be an answer. This could especially be true in certain kinds of aircraft of the military variety that might be exposed to high intensity electronic radiations and countermeasures. Using optical fibers to "sense" conditions on the aircraft—and these could be any kind of phenomena from heat to stress and strain, voltage and current, to pressures and thus altitudes—make them very attractive and useful in this application. Also modern military weaponry which may be exposed to the same type of environments will find optical fibers useful as a countermeasure.

LIGHT AND COLORS (MONOCHROMATIC LIGHT)

It is a nice experience to visit your local *Radio Shack store* or equivalent purveyor of electronic devices and examine the ICs, LEDs and optoisolator elements that have to do with light. We quickly find that LEDs are available in various color outputs such as yellow, red, and blue. We know that these LEDs are *monochromatic* because the word monochromatic means "one color." It also means "one wavelength" but we don't worry about that here as we identify a color with a wavelength in the visible light spectrum. For identification purposes, if we say a light is not monochromatic we then assume that the light has many colors. White light, for example, is composed of many colors.

In television work we know there are three primary colors, red, blue, and green. The manner in which these are mixed gives rise to a multitude of other colors, sufficient for us to say that we have "color" television. If you have ever adjusted your TV color controls, you know that the proper mixing of these colors will give you a grade of white ranging from grey to "white white." Thus, we might say that when we have nothing but blue, we have a *monochromaticity* and when we combine colors we have nonmonochromaticity.

Monochromaticity is important in the world of optical-fiber light-transmission, containment, handling, and the use of light in control, measurement, and sensing applications. If the light is monochromatic then the optical fiber can be designed so that it transmits that particular light best, and we don't care if other wavelengths are not transmitted or handled as well. In the visible light spectrum it is nice to glance at the end of a fiber (*obliquely* and *never directly* at the end for safety reasons—you *do* value your eyesight don't you?) and seeing the color you can say: "Well the wavelength of that light is about -------" and you fill in the appropriate *micrometer* dimensions. By the way, did you know that the U.S. Navy is very interested in optical fibers and their use and began a program in 1978 called *FOSS*—an acronym for *Fiber Optics Sensor Systems*. It is a big research and development program and most of it is classified. But knowing that this arm of our Department of Defense is vitally concerned with this material tells us how important fiberoptics will be in the future.

THE LASER APPLICATIONS OF FIBEROPTICS

Some years ago I performed some experiments with a ruby rod laser and had fun with it. A low power (10 watt) device, it was for instruction and demonstration of the laser's capability and components and construction. A small ellipsoidal chamber was filled with the ruby rod along one focal axis and a flash tube (xenon or similar) along the other focal axis. At first we satisfied ourselves with, (after mathematically describing the operation and so on) bursting balloons with the light ray, then, believe it or not, killing spiders (a myriad of which hung around the labs). Finally, I placed an ordinary metal razor blade at the focal point of the beam and pierced it with the ray. Of course, after this fun, more serious experimental work followed.

One of the nice things about our little unit was that it made a very loud "bang" when the discharge button was pressed, and that made all observers jump and regard the "black boxes" with much more respect. The bang was in part due to the "smacking home" of the large relay we used to control the very high voltage supply that, in turn, gave the flash to the "flash tube" in the ellipsoidal chamber. The inside of that chamber was burnished so brightly that almost any light used for inspection was hard to tolerate.

We got a pulse of pure red light that came out through one end of the ruby rod after bouncing from a mirror at the other end of the rod. The device was useful for demonstrations about 1962 or so. Later came the gaseous lasers which didn't pulse, but produced a continuous beam of light and along about this time came the solid-state lasers that were small semiconductor injection type diodes. They could be modulated at a very fast rate by other signals. These latter lasers, of course, are ideal for the injection of laser light beams, modulated or data handling or whatever, into optical fibers for transmission to another location, or for modification by some physical happening to the fibers so that what we call "sensing" of that outside physical phenomena might take place. We know now that laser applications include such things as "retina tacking." This is a medical process that uses a laser beam focused at precisely the right distance so that it can be used to repair a detached retina. As in our earlier experiment where we found it necessary to place the metal razor blade at precisely the focal point of the laser ray to "punch" a hole in the metal, so the fusing of the retina takes place only in a very thin focal plane of adjustment so other parts of the eye are not affected. Well, naturally, you don't "hunt around" trying to find the focal plane—you'd do damage then of course. The instrument in the hands of an experienced and qualified doctor, is pre-set to within millionths of an inch of the correct focal plane *before* energy is applied for the fusing. It is all very delicate and precise, but it is commonly done.

Considering the types of lasers, we might reject the ruby type and the gaseous types for optical fiber work because of their size and expense. The solid-state diode types are much easier to "interface" with the optical fiber or fibers and they can be modulated and can do almost all of the things required. They produce a polarized and monochromatic light and this is good. They are the type most under consideration at present.

This must not be interpreted as meaning that LEDs will not be used. They are undergoing improvement every day and there is now one which emits light near the infrared frequencies and this is appropriate to the use of such light in optical fibers. Also, LEDs of higher power outputs are now available and under development. Fiber-optic systems are currently in use in telephone work and in some television applications. The Department of Defense has some buildings which use optical fibers in their secure transmission and intercommunications systems. There is no radiation from them, so eavesdropping electromagnetically is impossible! Communications systems using optical fibers are numerous and expanding constantly.

LASER WEAPONRY

Recall our little experiment using a ruby laser to "punch" holes in metal razor blades? Well, of course the military establishments of the world are constantly trying to develop a laser of high enough power, small enough size, and economically feasible enough to use as a weapon or in a weapons system. Already laser light and PIN and avalanche-diode detectors have been used in precise "ranging systems." These determine the distance to an object by pulsing the laser light at it and timing the trip there and back, like radar does when it determines range. There have been reports which imply that laser weapons in outer space have been used to disrupt and destroy satellites in tests—highly secret of course—and we are informed that in space there is little or no attenuation of the laser beam signal.

Recalling our own experiments in which the focal distance to the razor blade was critical in order to punch a hole in it, one wonders if focusing of the laser beams of space vehicles has been made really possible. One day, perhaps, we shall know. But scientific evidence seems to indicate that only if that beam can be focused precisely (as for retina operations or razar blade hole punching) will the tremendous energy contained therein be available for death and destructive purposes. Otherwise, the light is there and in its modified and weaker existence can be used otherwise in many other applications such as transmission of data, communications, pictures, and so on, over fiber optical links and in some ways in military applications.

John J. Fialka, writing in the Wall Street Journal, has stated that in mock battle field games: "The weapons on both sides will be equipped with lasers; the firing of a blank round will send an intense, pencil-thin beam of infrared light at the enemy. The soldiers and vehicles will be equipped with button-sized electronic sensors that react when struck by the laser beam. A "kill" on a soldier will cause a steady beeping in a little box he carries; the sound of which can only be turned off by a little key attached to the soldier's rifle. When he uses the key the rifle no longer works. The soldier is out of action until the umpire sends a special electronic impulse that unjams the system. Since some soldiers and tanks will be equipped with *special* small black boxes which contain transponders (special devices which can be queried by a specially coded signal) these black boxes will record when weapons are fired and when and what kinds of hits are received. They will also send out position signals to some number of antennas surrounding the mock battlefield. A computer system will interrogate the black boxes 10 times per second, analyze the data, and give base commanders and technicians a running account of the battle." One might think of these laser weapons as *practice* weapons to improve one's ability to register hits on an enemy target. Such a system has never before been possible.

In a laboratory at the Johnson Space Center one scientist was recently concerned with running an electric toy train back and forth on about 300 feet of track. As the train moved, so did the bicycle-type reflector mounted on the engine and this reflected the pencil-thin beam of a surveyors-type laser instrument. The information from the reflected pulses was then used by a small computer to present a digital readout of the distance, at any time (t), to the engine. The purpose of all this experimentation was to develop a type of docking system—automatic and self guided—for use with satellites and other spacecraft. The laser, sending out the beam and thus producing the energy from which reflections from the "target" could be formed, *was not much larger than a grain of salt*—a gallium-aluminum-arsenide type of semiconductor. This laser emits light in the infrared region just past that region that is visible to the human eye. You cannot see that beam but it exists. Five tones modulate the light beam, and these are used as the basis for a comparison circuit to determine time of travel and position and such, so accurately—within millionths of a meter accuracy—to determine the target's location wherever it might be and whatever it might be, from a toy train engine to a spacecraft. But, as was indicated by the experimenter, in order for this type of system to work properly, there had to be some kind of auto-reflecting system of a kind of material which would reflect the laser light beam back to its source.

Radio shack makes an infrared, pulsed-light, announcing and security system that in many ways duplicates this concept if you wish to experiment. Furnished are the source of the light (laser) and a receiver in close proximity, and a special grating which can be positioned to reflect the infrared beam. The nice thing about this system is that ambient light from the sun or incandescent lights does not affect it, and you cannot see the beam—*but don't look into the laser lens to try to see it!* Instead of a computer which provides distance information, the system will sound an alarm if the beam is broken or interrupted.

SOME COMPARISONS BETWEEN COPPER
WIRE CHANNELS AND FIBER-OPTIC CHANNELS

Bell Northern Research Limited of Canada has done some study and made some comparisons among some methods of handling long-range communications of the telephone variety. These results can be of interest to us as we consider industrial applications of the fiber-optic communication and data handling systems. When considering the use of radio communications systems, the cost rose very high due to the necessity of having access points at about 16 kilometer distances along the routes. Also the system had to be bidirectional for two way conversations. Analog information types of systems picked up noise and garbled the speech and digital techniques were ruled out because of the non-availability of the coaxial cable carrier systems needed for this technique. With either analog or digital systems a large number of high cost repeater stations would be needed and susceptibility to ac interference would be very high.

In typical copper carrier systems up to 1344 voice systems or channels or digitally routed by copper lines to radio sites and at that point they are multiplexed so they can be handled by radio carrier systems. But, when considering the use of fiber-optic types of channeling and carriers and using a multiplexer at the nearby toll office, it was estimated *that only four fibers* (plus spares) would be necessary to handle all the channels of voice communication previously carried by up to 115 copper wire cable pairs (a decided difference in cost resulting). Recall that the optical fiber being considered here might be only 125 *microns* in diameter—and that is small indeed!

HANDLING LONG RUNS WITH OPTICAL FIBERS

Whenever distance is involved a consideration to losses and cable, fiber, or wire breakage must be of importance. The losses due to the transmission system are compensated for by repeaters and/or amplifiers of a suitable type. The breakages involve splicing and rejoining the ends of the wire, or fiber, or whatever. Note that when reamplifying the signals, the repeaters must be located at suitable distances so the losses won't get so large that the signal intelligence vanishes and can't be reamplified. You have to be able to detect the signals clearly to amplify them and send them on their way.

The splicing of cables and such may present a little different kind of problem. Splicing wires is easy. You just find the break, twist or mechanically fasten the ends of the broken line together then solder them for good electrical conductivity and strength and reinsulate the joint and you're in business! Mending a radio channel may involve replacing a circuit board in a transceiver or adjusting the antenna, or improving electrical connections in the system, or repairing the multiplexing computer or whatever—a somewhat more difficult job.

Repairing a fiber-optic line involves making a perpendicular end on the two sections, then bringing these together for a *perfect* pressure type fit, and keeping the two segments so connected for the balance of the time the system fiber is used. Considering that you are handling fibers of 300 micron diameter (with cladding) and the fitting together of the two ends, made perfectly parallel to each other, must be tightly and perfectly done, you've got a problem! Well, you've got a problem unless, as Bell-Northern did, you invent a machine to make that splice for you! Recall that an optical fiber cable may have just optical fibers, or it may also include some copper wire conductors for test, measurement, and such, and the cables themselves may be subjected to various chemicals which could act upon the cable material in a nondesired manner. There could be also some effect from nuclear radiations but this isn't likely. But any cable will be subjected to tension, twisting, crushing, and vibration and that is just a fact-of-life!

The cable must be designed for a minimum of loss. Bell-

Northern has accomplished less than 3 dB/km attenuation at 840 nm wavelength (nm is nanometer). This is a cable that has a flexible core of steel strands covered with a plastic that is grooved to accommodate the optical fibers and copper wire pairs. There are separate grooves for control, measurement, and test purposes. It is normal to use a dc current for control purposes but light rays have to be modulated and demodulated to produce the same selective controls. The cable must be wrapped with a type material that can withstand the environment whatever it may happen to be (a stretch of wilderness with extremes of cold and heat, a factory with chemical pollutants, a stretch of underground city street, with the changing pressures and stress and strains and so on).

SATISFYING THAT YEN FOR EXPERIMENTATION

Once we have sufficient knowledge of a subject to understand it from the theoretical viewpoint, we begin to desire to have a little hands-on experience to go along with that theory. If you now have that feeling, then look at Fig. 2-2. This shows a very simple system (using parts readily available from Radio Shack) with which you can do some experimentation with light communications and even light control systems according to your ability to invent, imagine, or design these circuits.

You will notice that the LED and receiver diode are Texas Instruments types but equivalent types can be obtained from Radio Shack. You will have to do some experimentation with the circuit shown. Also notice that the plastic optical fiber is obtainable from Edmund Scientific Co. They have various types of such cables that you can get to experiment with. For example, you might use a multi-cable of several strands of optical fibers, and then use various colors of LEDs to see how the different wavelengths propagate through the fibers. If you then figure out a way to pulse the LEDs with some form of code, you might design a control system so it can close some kind of relay (solid state or mechanical). It's a fun project for the dedicated experimenter.

Along with the propagation of light rays through optical fibers, you might want to experiment further with a system using lenses and fibers. Edmund Scientific has such a device called a fiberscope. It has coherent optical fibers with a (7×) magnifying eyepiece. This eyepiece focuses from infinity to as close as ½ inch.

USING LENSES WITH OPTICAL FIBERS

You may have already considered this and wondered why we

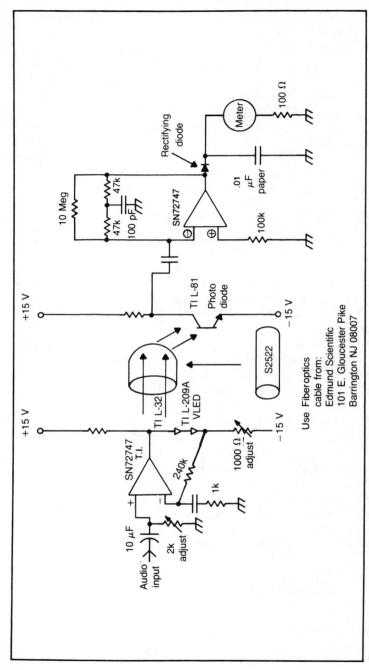

Fig. 2-2. A simple communications system using light beams.

54

have not mentioned it sooner. It just seems logical to use a focusing lens to get light into an optical fiber guide with as little loss in doing so as possible. Almost everyone has, at one time or another used a magnifying lens to focus the sun's rays upon something which burns—clothing, cigarette, wood shavings, paper, you name it! So we have some idea that the rays which impinge upon the side of the lens facing the sun, are somehow, gathered together and concentrated so that they increase in energy such that a tiny point of light becomes very bright and very hot on the other side of the lens. When we hold something at just the correct distance away—where we get the smallest point of light—the heat becomes so intense, that something tends to get hot and burn.

Next, we consider the size of the light source to the size of the optical fibers and we begin to see what could be a problem area. Recall that the optical fiber may be only some 50 to 130 microns in diameter (the core that is) and the larger the core the more tendency to have a "multimode" propagation inside that causes losses. We also note here that light rays may enter the fiber through what has been called a *numerical aperture* which is the sine of the angle associated with the cone of propagation into the fiber. For example, at Bell-Northern Research, light rays enter a modern type fiber within a cone which has a half-angle of 10 degrees. In this case, the numerical aperture is said to be sine 10°=.18. A large NA (numerical aperture) means larger signal, or ray loss, and larger distortion of the intelligence being thus conveyed.

So, we think of the size of, say, an LED and the size of an optical fiber and we think: "How the blazes shall we proceed to get the LED rays into the fiber optical light guide?" Examine Fig. 2-3 for a general illustration of this problem.

Of course, if we can "back-up" so that we have the integrated circuit structure of the LED or laser light source, and have some physical method of bringing the end of such a small fiber light guide *into* the acceptance cone angle for the guide, then perhaps we "have it made." The smaller the light source is physically, the better we will be able to capture the rays which emit from it.

When you do your experimentation, using, perhaps, an LED from Radio-Shack and an optical fiber from Edmund Scientific, you'll then be confronted with this problem. You'll probably just wiggle the fiber around, or the LED, or both, until you observe the glow at the fiber optical strand tip, away from the source, which proves that light is passing through it and emerging at the far end. To try to measure the losses you will have to use some method to determine the emitted energy at the LED surface and compare this with the emitted energy at the far end of the optical fiber. This measurement is a project all by itself.

We find ourselves back to the subject of lenses. If we can use the focusing ability of lenses to get light into the optical fiber, and then use some lenses to "magnify" the light coming back out of the fiber at the far end, we may better understand what is going on inside, and we may better use the fiber for whatever purposes we might have had in mind. We might consider two types of lenses or focusing methods. One might be to use a parabolic mirror *behind* the light source and place the input end of the fiber precisely at the focal point of the parabola as illustrated in Fig. 2-4. We have given some parabolic equation forms and the standard form for the type parabola shown. Note that the focal distance from the vertex to point (F) is always the same as the distance from the vertex line to the directrix line. In this illustration the focus point is 3.52 squares from the vertex and lies on the (x) axis. Any light rays coming into the parabola from the right (as we view it) would be reflected by this type mirror to the focal point. Thus, it tends to gather quite a large bit of energy (depending on its size) and it concentrates light at that one point. If your fiber-optic light source is broad-beamed, this type focusing element might be important. The parabola, of course, is three dimensional and in any direction from the vertex it has this shape as shown. It is, in a way, like a saucer or curved surface which tends to "cup" the received energy going into it.

The focusing of light using a reflecting mechanism is one method and the use of a lens is another method of focusing energy rays or waves. The lens works differently as light will pass through it but due to the bending of the rays they can be caused to hit a given point on the other side of the lens. When that happens, we say the light rays have been focused. Figure 2-5 illustrates this condition

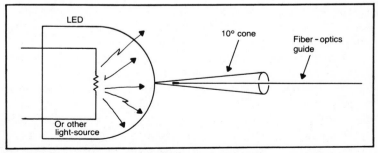

Fig. 2-3. Getting light into a fiber-optic lightguide.

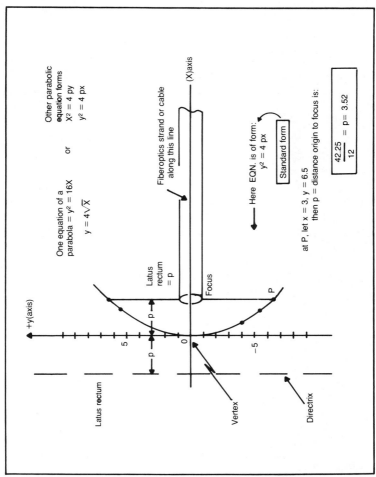

Fig. 2-4. A parabolic mirror can focus light rays the same as a parabolic antenna focuses radar waves.

Other parabolic equation forms

$X^2 = 4 py$

or

$y^2 = 4 px$

One equation of a parabola = $y^2 = 16X$

$y = 4\sqrt{X}$

Fiberoptics strand or cable along this line

Here EQN. is of form:

$y^2 = 4 px$

Standard form

at P, let x = 3, y = 6.5
then p = distance origin to focus is:

$\dfrac{42.25}{12} = p = 3.52$

(X)axis

Focus

Latus rectum = p

P

+y(axis)

5

0

-5

Latus rectum

Vertex

Directrix

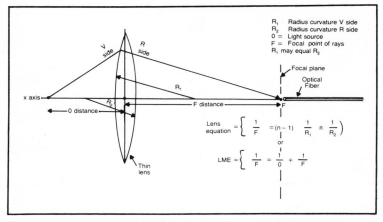

Fig. 2-5. A focusing-magnifying lens and the "lens-makers" equation.

for the so-called "thin-lens" type of magnifying glass. We assume here that the lens is so symmetrical (designed that it makes no difference which side you present to the light source and which side you present to the optical fiber end).

USE OF A SPHERICAL MIRROR FOR FOCUSING LIGHT RAYS

Figure 2-6 shows one example of how a small spherical mirror might be used to focus the rays of light from an object at (0) to the focal position at (F) where the end of our fiberoptics might be located. It is possible to determine the measurements and distances involved by using the equation shown. Just be sure to use the same units (and we suggest metric) to find the value of (r) or the value of (F) and (0).

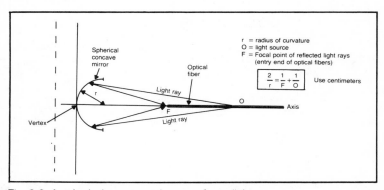

Fig. 2-6. A spherical concave mirror can focus light.

Normally, when you have an expression in three unknowns you have to eliminate some of them to get a solution. So, how do we do this? One simple way is to *select* a small spherical "thing" which has good reflective qualities and is shaped as shown, measure the radius (r) and you have one of the necessary values. Second, you might *specify* a distance from the vertex to the light source (0) and then solve for the remaining unknown, the distance from the vertex to (F). This is perhaps not the best way but is a method that should work. In any event you will get some focusing of the light rays into your fiber-optic "light guide" if the end of the fiber is very accurately positioned at (F) and the light is properly located precisely at (0) and the mirror reflector has good reflecting qualities for the light frequency being used. But experiment! It is fun and when you succeed in such an experiment, the pleasure is very large. Remember, small errors in measurement or positioning can render the experiment useless.

THE INTERFACING OF LIGHT SOURCES
AND RECEIVERS WITH OPTICAL FIBERS

In the experiments just described, you have considered the real basic problem. How do you use large size light sources and reflectors or lenses to get that light into such a tiny strand of special glass which makes up the core of the optical fiber, and do it with any efficiency at all? One answer might be to reconsider some things that may not be available to us and that is the extremely small light sources and diode receptors which can be connected to the ends of optical fibers using special splicing and interfacing machines. Once we are able to make these connections and have these kinds of devices, then transmission of intelligence, data, or whatever using light rays is not a serious problem. (Someone will surely disagree with that statement!)

We shall find out that being able to properly code the data so that it won't be lost due to delays in propagation and the changing shapes of pulses and so on, becomes a problem instead. Also the reinforcement of the lost energy in transmission is a requirement and as distances involved are relatively short before reinforcement is required, this may become a problem. Lifetimes of components, accuracy of stability of the frequency of the light rays, and so on must be considered.

So, how do we interface two optical fibers so that the light rays pass through them? Bell-Northern Research has come up with a simple method which is effective in the field. The optical fibers are joined together by a fusion process which is made possible through the use of a special splicing machine developed just for this purpose. What happens is that when it is necessary to splice an optical fiber cable, the protective coating is removed. The ends of both sides of the splice are cut perpendicular to the length so that they will join together perfectly. Next, they are placed in the special machine, but separated slightly. A small arc softens the end of one fiber and rounds it slightly to insure that the material will fuse outward from the core in the actual fusion process. Then the two ends are brought together inside of the splicing machine and an arc that is hotter makes the glass in each half flow together to form the splice. Much care is taken to prevent air bubbles forming at the splice interface. The fiber is then rewrapped and placed in its cable package and it is again ready for use, or is ready to send the light rays through this joint on the next leg of their optical-fiber-guide journey. Sometimes, due to the fact that losses will be increased by such splicing, reinforcement of signals will be necessary. At the end of a fiber-optic line, it is common to use a cylindrical plastic or lead "head" that surrounds the fiber and thus protects it and (if it is in a cable) the end of the cable may also have a protective element of this type covering it.

The interfacing of a receiver or light source with an optical fiber may be done mechanically, or it may be done by properly bringing together the fiber and the element, and in a sense, fusing them into one unit. Also, a "jacketed" optical fiber may be used to connect an optical fiber to the card edge optical connector from a laser. From the card, a distribution system can be fabricated so that the cable connection to the optical-fiber containing the light ray and intelligence can be accomplished in a multiple connecting arrangement.

What we are saying is that if you bring in the light and intelligence on one fiber, there are physical interfacing units that can be used that can connect this one optical fiber to a multitude of other fibers and thus send that intelligence over many "lines." Such is a desirable case if you are using the light ray for telephonic communications.

SPLICING OPTICAL FIBERS

Three methods of splicing optical fibers are shown in Fig. 2-7. At the top (A) we see the "arc" method previously described in the text. A special machine prepares the ends of the fiber, supplies the arc and then brings the somewhat plastic ends together for bonding

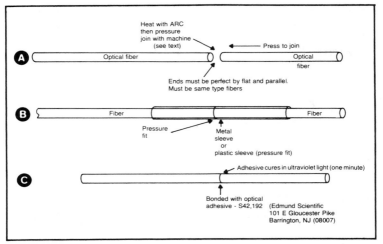

Fig. 2-7. Methods of splicing optical fibers.

molecularly. A strengthening jacket or sleeve will cover the splice. In the second illustration (B) you just have a pressure bonding of two fibers whose ends have been carefully cut so they are exactly flat and parallel. Then a metal jacket covers the cladding and plastic covering of the fiber and hold the two ends together. This is not as good as other methods as strain may separate the fibers. In the third illustration (C) an "optical adhesive" glue is used to bond the ends of the fibers together, and then the metal jack covers the splice to give strength to the joint.

The following illustrations show (in a very basic sense) how you would make an optical fiber transmitter and receiver. See Fig. 2-8. A laser transmitter is somewhat more complicated. The laser must be cooled or at least maintained within a certain temperature range to emit the right kind of coherent light. Another simple diagram of a fiber-optic communications system is shown in Fig. 2-9.

The *LED driver* is that unit which causes the output of the LED to vary or cut on and off at some rate equal or proportional to the combined input signals that are to be converted to light pulses or variations and so then transmitted over the optical fiber channel. Remember, that one of the advantages of using the fiber-optic channel is that it will be secure from electrical interference and eavesdropping. Multiplexing means combining of signals together so they are transmitted as a complex modulation—like a flute, harp, piano, and horn are all transmitted by radio or TV (or as a time-

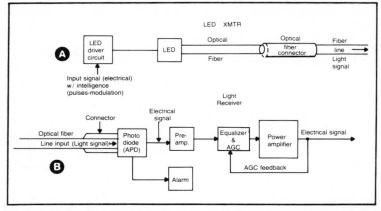

Fig. 2-8. A light-transmitter block diagram (A). A light-receiver block diagram (B).

sharing system in which each input is sampled for a few fractions of a second and that information is then transmitted, so all lines are sampled in sequence). The sampling must be very fast and this requires a somewhat complex electronic switching circuit to do the

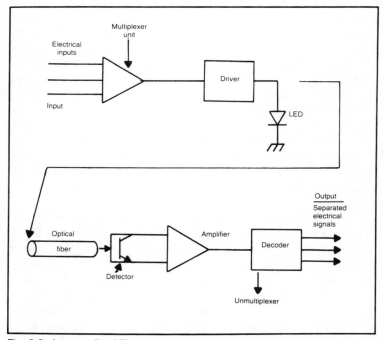

Fig. 2-9. A generalized fiber-optic communications system.

job if all information on each line is to be obtained and sent along as a light ray. Also the LED must be able to follow the very fast sampling units. The *decoder* (or unmultiplexer for lack of a better word) separates these frequency units or time units (or combines the time units as the case may require) so that the originals of the inputs are preserved at the output. Notice we have used some symbology for the LED and photodetector which are very common in practice and relate to diagrams of photo-optics you may find anywhere.

It has been found that the output power of a laser light source will vary more than an LED *when the laser is subjected to temperature changes*. This means that a bit of circuitry that will monitor the temperature (requiring a temperature sensor and an output sensor) must be so connected that it will cause a proper feedback to stabilize the lasers output. One example is shown in Fig. 2-10.

SOME WEIGHT CONSIDERATIONS IN COMMERCIAL APPLICATIONS

We are aware of the weight of copper wires. If we have miles and miles of lines, the weight is extremely large. We do have miles and miles of lines necessary in modern aircraft, for example, and it would be nice to reduce this weight so the airplane could be more cost effective in the many ways in which cost effectiveness is evaluated. It has been stated that 1.5 pounds of fiberoptics can replace as much as 30 pounds of copper wire in an airplane of the *A7*

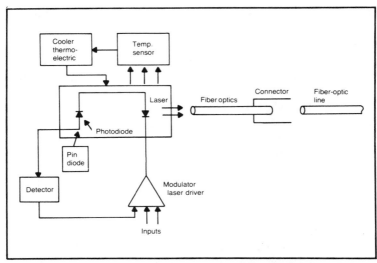

Fig. 2-10. Stabilizing a laser light source transmitter by using temperature control.

type. The electromagnetic field associated with the passing of current through the copper wires is not present when fiberoptics are used. This gives a quieter atmosphere for the complex instrumentation in many types of airplanes.

At the same time, one of the principal disadvantages of using fiber-optic cables is that they cannot carry direct current. And this type current is commonly used for control and switching applications as well as for power to the various units (amplifiers, equalizers, and sensors) that are part of the optical fiber systems. The way many companies get around this problem now is *to include* a few pair of copper wires in a fiber-optic cable sheath which may have many optical fiber strands or lines also contained therein.

THE OPTOCOUPLER

For experimental purposes one can obtain from Radio Shack an optocoupler with a linear op-amp output. Because such a device assists us in our understanding of the use of light as a transmission channel, we examine it briefly here. The optocoupler is an integrated circuit unit, and it is very small, about ⅜ inch by ½ inch by ⅛ inch in size with three prongs extending from each side. Inside of this hard plastic package are the light-emitting source, the receiving diode element, and a small operational amplifier. The amplifier requires 12 volts dc at 13 mA. The input light diode uses 3.0 volts at around 50 mA.

The operation is basic. You apply voltages and current to the input diode and it produces light rays inside the structure that go through space to the receiver diode element. These are then converted by that receiving diode into electrical impulses whose magnitude is dependent upon the intensity of the light impacting the diode element. The receiving diode is then connected to an operational amplifier which has the capability of amplifying direct currents and providing an output through one of the pins that is exactly proportional to the output of the receiver diode's output. We can then use this signal for various purposes. In Fig. 2-11 we see how the connections to this type unit are made.

What is interesting to us about this unit is that the infrared light rays are emitted from the gallium-arsenide diode and they are coupled physically through space to the receiver diode without any light-guiding path necessary. The diode emits a beam which (for a short distance) will provide enough intensity upon the receiver diode so that the light-coupling takes place. We imagine that we can use such an element coupled into an optical fiber in such a manner

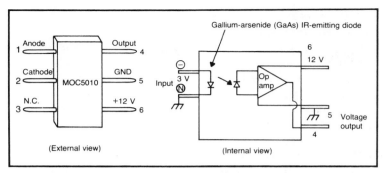

Fig. 2-11. An optocoupler with a linear operational amplifier.

that its light beam goes down that fiber for a much longer distance before it terminates right upon a receiver diode such as this, and so reconverts the light rays back to electrical signals. The amplifier will be required to amplify the signal to a usable level.

If the output of the op amp is routed into a circuit which will provide a *current* output exactly proportional to the voltage output of the opamp that we might use this current to cause another GaAs diode to emit light rays. If it is coupled to another optical fiber we have the equivalent of a repeating unit (on a small scale) such as is used in telephone communications with optical fibers.

It is current practice to use optoisolators when connecting a computer to outside machines. Although the path between the light emitter and the light receiver is very small, it is an air path or space path through which no electrical connection is made. This means a complete isolation of the light-emitter side of the circuit from the output-receiver side of the circuit. Thus, any stray voltages, static, or malfunctions that happen in the output circuitry are not coupled back into the input (and into the computers). If the input side of the circuit is concerned with very small voltages, then one can use these to control larger currents and voltages in the output by activating such devices as a *triac* (voltage controlled switch) with the current from a receiver diode (amplified as necessary by some integrated circuit transistor or transistors in a *Darlington-type* arrangement). You can find such devices for sale in radio parts stores as an integrated circuit of very small size, but with large capability.

What all this proves is that *control* functions can be accomplished using a light ray path to convey the control information. We must be aware that this signalling is relatively slow and so the light emitter and receiver can be relative slow in response time. As we consider the use of transmitter and receiver elements for use with fiber optical systems, we must think in terms of a very fast, turn-on and turn-off condition for light. If it transmits fast pulses using the binary code, it must have just as fast a response from the receiver so that the pulses will be reproduced with as sharp edges as possible even though they are being reproduced at a very fast rate because they were so transmitted. We are aware that up to 50 megabits/second might be the digital rate used in some applications.

SOME OPTICAL MIXING IDEAS

Basically, all we have to do is to take the light emitted from one optical fiber and reflect it into many fibers in order to get a "distribution system." A special coupling device can be used for this purpose that has the proper size and mirror characteristics to provide the reflections in the proper directions. Figure 2-12 might help us visualize this kind of unit. The whole unit is encased and the mirror must be specially shaped and positioned to get the maximum return of light energy into the multi-strand optical fibers in the block. As you readily see from this illustration, the information contained on the input light rays is now distributed to many channels by this method.

The mixing of signals means that the optical-fiber system must be able to mix different light rays. Perhaps we can state this more clearly by saying that in a mixing process the variations of light intensity, that may be from dark to full brilliance or any shade

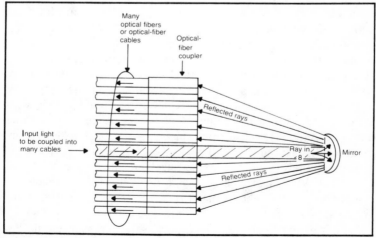

Fig. 2-12. A multi-strand coupler for optical rays.

between, must then by imposed on other rays going to other channels. Now, perhaps we find that some electronics might be required to do this job.

If we take a light ray from a fiber-optics line and pass it through an electronic circuit which will convert the variations in light intensity into a proportional electrical signal, and we amplify that signal and use it to modulate an existing beam of light, we can say that we have "mixed" the signals on the original beam with those from the demodulated beam of light.

We know that there are some physical devices that can be caused to vary their transparency to light rays if an applied voltage is placed across them. Some crystals have this property, and so they might be used as modulators for a light beam that will be the "carrier" of the mixed signals. Then one might have a separate beam and take all the signals that are to be mixed (that come from other beams or rays of light) and convert them into electrical signals, mix the electrical signals so that just one complex electrical signal results, and then use that signal to modulate the freshly generated light-ray carrier. Demodulation requires a receiver, a light-to-electrical-current diode and amplifier, and some demodulators or filters to again separate those electrical signals back into each individual channel.

FURTHER USES OF LIGHT RAYS AND OPTICS

We have mentioned "ranging" and surveying using light rays that can be flashed to a distant point, reflected and timed, so that distance can be easily calculated as it is with radar. Unlike radar, it is said that light rays are not vulnerable to *ECM* (electronic countermeasures) or nuclear residue in the atmosphere and so on. But, one wonders whether or not the actual effect of atmospherics might not be a prime consideration in the use of light rays and optical devices. Perhaps you have been one who has traveled the deserts of the world, or the U.S., or even looked along a hot highway at midsummer and have seen a shimmering scene. The reflected light rays that make up our vision are certainly affected by the rising heat waves. Sometimes a "lens" effect is present which tends to magnify or reduce the scenes around us. So, although light rays are better than some types of electronics they may be susceptible to other effects which are bad. We need to be aware that this is possible.

Some sensors operate on the existence or absence or variations in light intensity. It is a simple task to provide a rotating wheel with a small mirror or polished surface that will reflect a beam of light into a "receiver cell" once each revolution of the wheel. A digital counter can easily add the pulses of electricity that are the converted light pulses, and give us speed of rotation (by stating the pulses-per-second or the total number of revolutions, which might specify a distance if the wheel is on the ground). Using optical fibers to conduct the light rays to the wheel and then conduct the return flash-of-light to a conversion diode enables a designer to locate the sensitive equipment, and computing the equipment, where it will function best.

Light rays can be used to determine colors, and this could be a boon to manufacturers who want a machine which can detect and separate various items that are color coded. It is a well known fact, for example, that light rays going through a lens filter can be restricted or enhanced by that filter. A green filter may pass or inhibit more of the rays in that part of the spectrum than it will other colors. Thus, with a filter, a big output in reflected light from a sensor could mean a "green" object is present in the view of the light source and the receiver. But, also, a restrictive green filter might discriminate against the green light rays and so anything "green" or producing "green light" will not produce a big electrical signal, but the smallest possible electrical signal, or even no signal at all!

We know there are green, red, ultraviolet, blue, yellow and other filters used by camera buffs. Adaptation of these into the world of "sensory optics" is already an accomplishment. You can easily conduct an experiment which brings home this idea of the magnitude and type of reflected light. Just use your camera lenses, or a light source and a light meter and measure the reflected light from various sheets of the so-called "construction paper" available in most grocery and pharmacy stores. These are rather large sheets, about 18 by 24 inches or so. Try white and green and black and red and orange and yellow and such. Always have your light source exactly in the same position with respect to each sheet, and the diode-type detector circuit or meter located in the same position relative to each sheet. If you use your SLR camera, position it in the same position with respect to each sheet. Then make a graph of the meter readings and you can determine the changes in reflectivity for the various surface colors.

For the use this idea in other than camera work, you need a diode-type sensor which will be sensitive to a certain color of light. Try Radio-Shack and find some diodes that have the largest output with red, yellow, green, or orange light. Connect your unit so that its electrical output is amplified by a small operational amplifier to a value of 5 or so volts and use that with some transistors as amplifiers (current types) to operate some small relay of the reed or moving pole type. You can use that to control larger currents.

Another way in which a light-emitting diode and a light-sensing diode are used is to place them opposite each other and then they can detect the passage of anything between them. The pulsed light beam announcer is something of this form, and these units are available through most radio-parts stores.

The "close proximity diode light source and receiver" have a most important application in a shaft encoder and decoder system. In this context the light is on one side of a disc and the receiver diode on the opposite side. The disc has holes or slits in it. If the disc rotates from a given position to the right, say, then counting the pulses tells us the position of the shaft. This will be very precise if we have a hole, say, every 10 degrees around the edge of the disc. A computer will keep track of how many pulses are accumulated in a forward direction as the shaft is rotated to move an arm, a tool, or whatever. The computer can then reverse the process by causing the shaft to rotate backward the same number of pulses, or stop at any fractional number of pulses between the maximum number and zero pulses. Thus we can "sense" shaft positions by a very tiny optical encoder-decoder system and do this more accurately using light rays, than we could using electrical currents in selsyns or potentiometers. At least there would be no physical connections or such limitations on the optical encoding-decoding system as far as rotation is concerned. There are actual physical limits to how far you can rotate a selsyn or synchro before a phase repeat takes place. There are also physical limitation with potentiometers.

Using light rays and optical fibers we can see into machines "down-deep" where we couldn't get before without disassembling them. We can see what is going on, how parts are wearing, or if things need oil, or are broken etc. Optical fibers can bring out visual signals of correct or incorrect operation to a dashboard of an automobile.

We must not overlook the use of light rays to measure the level of fluids. Some commercial applications use this method. It is simple in concept. You just place a light source on one side of a transparent vessel (glass) so the light rays can shine through when there is no fluid at the light level. Then place a photodiode that is sensitive to that light color on the other side of the glass container. Connect it through an amplifier to a control valve that is electrically operated. When the fluid is high in the container it will reduce the light and even prevent it from passing through the vessel. The photodiode then receives no light and produces a no output. If the fluid level drops (due to usage or whatever) then the screening effect of the fluid to light vanishes and the light rays pass through the vessel and hit the photodiode causing a voltage output signal that can cause the control valve to operate and refill the vessel until the liquid is again between the light source and the photodiode. It is like your commode but instead of using an air float to turn on and off the water you use light rays and electrical signals and an electrically operated fluid control valve. You'll think of many variations and improvements to this idea.

THE BASIC CONTROL ARRANGEMENT
USING LIGHT RAYS AND OPTICAL FIBERS

Whether it be for a robot (see our books, robotics TAB book numbers 1071 and 1421) or for steering some machinery, positioning something, control arrangement or identifying something by its presence or absence, the basic control arrangement is shown in Fig. 2-13.

The idea, again, is simple, you just position an LED or other light source (it is best if you use a light color which matches your photodiodes so you get the maximum output from the reflections). This means that if your photodiode has its best and highest output with a red color (infrared) then use a LED which produces that color as your light source. You can use white light (which has all colors) and other colors, green, blue, violet, yellow and so on, if you match-up your source and receivers.

Position the light source so it will shine on the object, path, line, or whatever, and position your photodiodes so they can get the reflections from that object. When the reflections are equal in intensity, as determined by the comparison amplifier output, that then will be zero or near that voltage—nothing happens. But, if the object shifts so that the reflected light is stronger on one photodiode than the other, then that diode will produce more output, and the output from the comparison amplifier will be either positive or negative, depending on how you've arranged your circuit. This can be used to state the direction something must move to reestablish balance in reflections. The amount of the comparison amplifier's output can tell the machinery how far to move or how far and how fast to move to keep up with the object.

Using this method you can "track" a line, cause a remotely controlled car to follow a line on the floor, cause a robot's gripper to

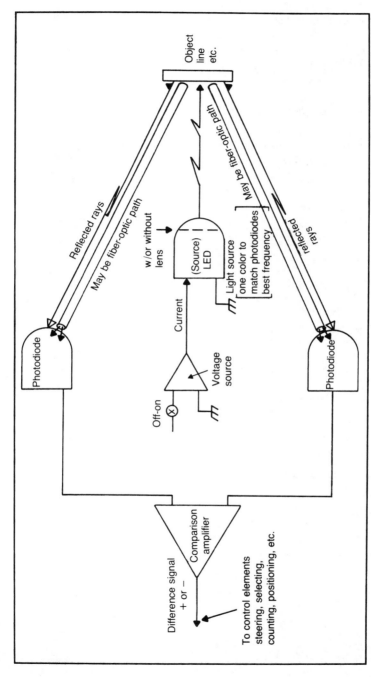

Fig. 2-13. Using light for control purposes.

close on an object, or search for an object. *If* you want to put your light source on a moving object and your photodiodes on another object capable of moving then the second object (car or whatever) can be caused to follow the first by moving so that its light reception is equal in its two separated photodiodes.

COMMERCIAL MANUFACTURERS OF LIGHT DEVICES

Once we get into the possibilities of using light for various purposes (and remember it can be rays that are visible or invisible from standard light sources, lasers, or LEDs, and since we have some ideas as to how to connect photodiodes so we can get voltage or current outputs proportional to the intensity of the light falling on them) we have the "makings" for lots of fun and profitable systems. Sometimes we need optical devices and Oriel Corporation is one

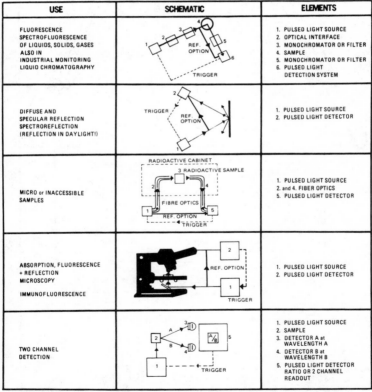

Fig. 2-14. Some optical systems you can assemble from standard parts (courtesy of Oriel Corporation).

manufacturer which makes a lot of these. They have a handsome catalog which shows what they are and how to use them. In Fig. 2-14 we show some ideas which come to us from the Oriel Corporation of Stamford Connecticut 06902.

When considering lasers, there is another firm, Lumonics, that has done much work in this field. We shall go into lasers in more detail in further pages. At the moment, however, such firms as Texas Instruments, have produced small laser units (GaAs) that work well with fiber optical systems. Lumonics produces CO_2 and multi-gas pulsed lasers of higher powers.

Chapter 3
Manufacture and Use of Optical Fibers

It is almost self-evident that we must have some kind of input-output device to use with optical fibers. Such equipment should enable the user to use his existing electronics with this relatively new means of the transmission of signals. We can't say that this is a very "new" method for as far back as Thomas Edison's time some experiments in the use of light rays to convey intelligence were being conducted. Also, Alexander G. Bell did some work with optical systems in his "Photophone" experiments in 1880. It wasn't until the 1900s that development of the actual optical-fiber "light guide" and analysis and development of the theory of how it worked enabled scientists to come up with useful devices such as the fiberscope (endoscope) for inspection dark and dangerous places.

THE TELEPHONIC COMMUNICATIONS PROBLEMS

We must now consider some future aspects of fiberoptics in that (1) they do exist (2) they are most advantageous (3) they can enable us to do some things better than we have done them before (easier and more economically) (4) they are still and will be for a while under more research and development. Let us consider the enormous problem involved in telephonic communications (not just the voice part but also the data transmission requirements and the video requirements and such).

When we want to really flabbergast ourselves, let us sit down and just try to imagine what it takes to connect every human being in the civilized world to every other by means of a reliable and reputable and consistent communications system. Further, if that doesn't mind boggle us, let us further imagine what it takes to be able to permit all these homosapiens to talk to each other at the same time.

Perhaps everyone has seen telephone crews "down in the hole" or "up on the pole" carefully splicing the countless numbers of specially color-coded cables together so that communication can be established, maintained, or continued to some area. As we visualize the birth rate of the world and the inevitable increase in demand for more and more communications, data, monitoring, signaling etc., we just can't imagine how they can keep up with the demand.

From our previous discussions you know that a few pounds of optical-fiber strands can replace countless pounds of copper-wire cables, so you should have no problem visualizing that gradually but inevitably the telephone cables will be replaced with these kinds of connecting communications channels. That part is not hard to imagine. But what we must also imagine is: How do they connect to existing equipment, designed for use with copper cables and to telephones and TV cables and such, without redesigning such equipment, replacing it, or modifying it? We ask ourselves: is it necessary to replace and redesign such equipment? We give ourselves an answer, saying: If we can use proper input-output devices which will *"interface"* (connect properly) to the new type optical-fiber cables and existing equipment, then we can continue to use existing equipment until something better to replace it comes along.

We also imagine the use of such devices in the field of computers. Already we are aware of the tremendous advantages that might be gained by the use of optical-fiber cables between main-frame equipment and terminals and such. Don't you recall these advantages? Well, reduction or *elimination* of induced radio-frequency interference, the ability to go in some places wire cables cannot go due to chemical, temperature, or other problems increase in data handling capability (even perhaps as high as ten-thousand fold) less loss in transmission and so on are some of them. We may find that the computer will need optical communications to speed up the data handling operations. There are many scientists who say that the travel of electrons over copper, silver, or gold plated wires is far too slow for the number of operations, calculations, and manipulations currently required of computers. We haven't even approached the ultimate goal of speed of computer operations in magnitude of operations performed per second or speed of operations performed per second. This goal is often stated as being the equal to that of the human mind!

THE SIGNALS

You can't interface anything with anything until you understand what the signal transference, conversion, transformations or whatever might involve. In order to know this you must know what kinds of signals are put onto a line (of whatever type) and then you design around the handling of those signals on the type of line or lines used.

Basically there are two types of signals used over lines of the types we are considering (wire or optical fibers). These are pulses and continuous wave signals. The pulses can come in various groupings and codings. They can be sent very fast and they enable a large number of channels when they are "time multiplexed." That is, when they are transmitted for a certain time duration from one channel input then the input channel is switched to another for a short time duration and then to another and so on until all input channels have been *sampled* a certain number of times per second. Each group of pulses transmitted over the connecting line will then be a certain amount of intelligence from a particular line. These pulse trains or groups are identified by some coding means so that they can then be channeled into the corresponding output channels and thus they carry all the essential information from the input line to the output line. Figure 3-1 may help you to understand this concept.

In Fig. 3-1 you see how a signal might appear coming over a line when it is time multiplexed (A). Each little group of pulses will have some identification arrangement so that the receiving equipment can tell it belongs to a certain group. The receiving equipment will then "add-up" the similarly coded pulse trains and put them together to make up the intelligence information (video, binary data, communications, control signals, etc.). Notice that at F all inputs go to a "*strobing*" amplifier which is then connected to a single line at its output. Thus each group of pulses having whatever kind of coding, such as at B, C, D, E, or some other arrangement, will be transmitted and separated timewise during the transmission. So, you ask, how can intelligent information be gleaned from "snatches" of data or just samples of information presented at the input? Shannon, of Bell Telephone Labs came up with an answer to that in addition to the one we provide here—that is *if you sample fast enough*, at such a high rate compared to the intelligence changes,

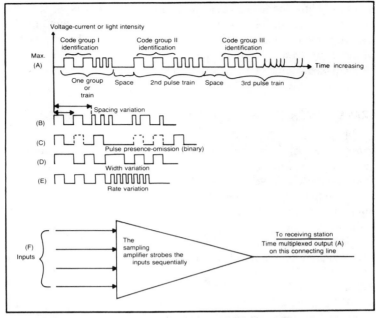

Fig. 3-1. Illustrating the transmission of pulse-trains or groups of pulses.

that you will never know that the information you get is being sampled at all! Some things like video pulse signals and computer pulse signals may present a different problem because you may not be able to "drop" even a single pulse from the input line without causing some *bad* change in the data output or recovery. But, for things such as telephone conversations which are basically continuous wave type transmissions, sampling of these kinds of signals (if done often enough) can result in recovery of the same type sine wave in the output (by filling in gaps by knowing how the signals are varying in amplitude because of the sampling) without loss of intelligence.

That was what Shannon proved in his experiments and studies and what he stated in his writings. That we humans use a lot of "noise" that we could omit and still get across to others the exact meaning of what we want to convey to them. Listen to a conversation and see how many sounds or syllables might be omitted and still not cause a loss of intelligence! Your mind follows the line of thought and so it tends to "fill-in" any blank spaces in actual communication. Shannon showed that in transmitting some signals (such as Morse code) one might miss a letter here and there and the mind would fill-in the gaps to make up the proper words. So it could be with our transmission of intelligence. Our I/O equipment might not send some signals properly, or we might miss a few fractional sections by time sampling, but the receiving equipment (input) could make up what is missing from a knowledge of what was going on before.

TRANSMISSION OF LIGHT SIGNALS (THE DISPERSION PROBLEM)

If we have an optical fiber which is single mode type (propagates the signal evenly along its length) then we may not have a bad problem. We will have losses but we may not have bad pulse distortion. But, if we have a cable or optical fiber which can be multi-mode in operation (and we tend to think that any optical fiber might go into this type operation if the frequency of input gets high enough) then we can have some elements of the signal such as the frequency components traveling through the fiber at a faster rate than other elements and then we do have a dispersion-distortion problem.

Everyone knows that when transmitting pulses you must have a broad bandwidth or you tend to round-off the pulses. The sharp corners of the pulses require high frequencies. Remember how you used to test your audio equipment by sending square wave pulses into it and then evaluating the shape of the audio pulses at the output? If they were nice and square and like the input pulses, the audio amplifiers were excellent in high frequency response. If the pulses were rounded off, then the frequency response was not so good. The same idea applies in fiberoptics. If we lose some of the higher frequencies being transmitted due to dispersion and multi-mode operation (recall that multi-mode means multiple path traveling of the various frequencies and subsequent phase differences that when recombined might cause reinforcement or cancellation of those frequencies) then we have pulses which might almost vanish. It has been found that by "grading" the index of refraction in the optical fiber material, a kind of "*monomode*" or single mode transmission of the light ray signals can be accomplished. This way the pulses are preserved in shape and number and the intelligence is faithfully transmitted. Again, we have to remember that it has been said that the transmission of light rays is just a very high frequency transmission of electromagnetic waves and so the equations and solutions to the light problem can follow the techniques used for, say, radar waves of very high microwave frequencies in following metallic or dielectric guides for these type of waves. There are many texts which analyze the transverse electric waves and the transverse magnetic waves associated with the propagation of these radar frequencies so if you're inclined toward some advanced mathematics we suggest you pursue those as a beginning study.

To round off the idea we presented previously, recall that in testing our audio equipment with square-wave pulses we were told to be happy if we had good flat topped pulses. That meant that the low-frequency response was good. Thus, if we built an amplifier and tested its response with square-wave pulses at the input (with a scope and speaker on the output) and we got good *square* pulses out and maintained these as we varied the input frequency from zero to, say 10 to 20 kHz we could be very proud indeed. Then, of course, we added equalizers and tone controls and such to make it sound like we needed it to sound to be pleasing to our ears!

TESTING OPTICAL FIBERS FOR TRANSMISSION CAPABILITY

It would seem logical to use square-wave pulses of light into optical fibers to test them. One could determine by the shape of the recovered pulses at the output whether the fiber was operating in the single or multiple mode type operation by the shape of the pulses. If the effect of the dispersion were large, a group of pulses might appear as shown in Fig. 3-2.

In free space light rays, like all other electromagnetic waves, travel at 300,000,000 meters/second (186,000 miles-per-second). In the type material used for optical fibers (lithium niobate, for example, which is called *birefringent*, meaning that a ray of light will divide forming two images) the beam will divide into two beams which are polarized with their planes of vibration at right angles to each other. This division occurs when the original light ray is passed through certain crystals such as calcite, quartz, ice, and

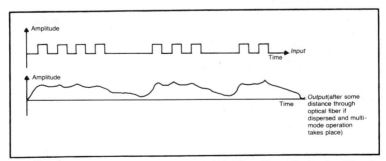

Fig. 3-2. The distortion of light pulses due to inter-fiber dispersion and multimode channeling.

dolomite. The speed of propagation through the material is different for the two rays and so they tend to "doubly refract" or have a double bending when passing through the crystals. Huygens discusses the property of *birefringence* in his *Opera Reliqua*.

What the above means to us is that the light rays may travel at different velocities through the optical-fiber material if they are not coherent and monochromatic. That means they must have the same phase (as they have when coming from a laser and having the same color or frequency that is also a laser characteristic). This means that we must be careful of our input device (the I of the I/O) that it presents to the optical fiber the right kind of light ray for accurate passage through the fiber material and has a minimum of dispersion and attenuation due to that passage. Laser light is one answer and perhaps some LEDs can be made to provide another if they are filtered or generate only a given light color(frequency).

SOME POLARIZED LIGHT EFFECTS

We won't get much of a "handle" on the concept of polarized light (having the electromagnetic vector, the (E) vector, in some plane *for one half of a light ray* and at right angles to it *for the second half of the ray*) if all we have to use is our imagination. Few of us can get the necessary *Nicol prism* to experiment with. But we can get some feel for light polarization by spending a few dollars to get a set of polarized sunglasses.

I have such a pair of sunglasses and enjoy riding down the highway and then twisting my head slightly and watching the effect on the reflected light. When they are positioned properly, the light reflections are almost cancelled out. A rotation of the lens permits more of the randomly polarized light rays reflected from the highway to pass through them and we see light reflections and more brightness—try it!

I recall a college physics experiment in which a grating was used in front of a light. When this grating was turned so that it permitted only the passage of certain polarizations of the light you could see the light. When it didn't permit the passage because it was polarized at right angles to the light rays you did not see the light behind it. Fascinating! So it is with the birefringement concept (or double-refraction concept). The light ray tends to divide inside the crystal and become two rays emerging from the other side. Thus if you looked at something from the *"two-beam"* side then light reflections in the single beam entering the crystal would be divided and

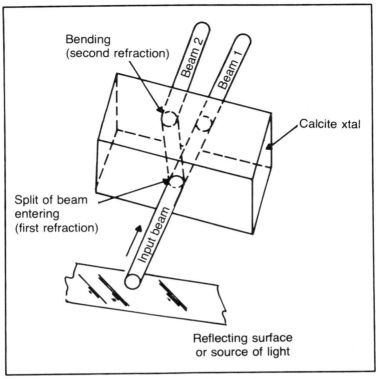

Fig. 3-3. The double beam output of a calcite crystal.

you'd see a double image of the subject material. Figure 3-3 will clarify this idea.

In the commercial world it would be relatively easy to "chop" a light source by having the light source polarized in a given direction (as from a laser) then passing this light through a polarized grating that permits the light to pass when the gratings match the light's polarization and cause the light *not to pass* when the grating is at 90 degrees to the plane of light polarization. Rapidly "flipping" the grating or causing it to change its polarization rapidly when a voltage or current is applied could be a method of modulating the light rays effectively and produce an off-on switching effect.

OPTICAL TRANSMISSION WINDOWS

It has been found by mathematical treatment and analysis and also by experimentation that there are what are called "optical windows" in various materials. What this means is that at certain frequencies the waves will pass through more readily than they will at other frequencies (of light). It seems that for optical fibers, if we have some frequencies above 1.28 or so micrometers (or in the infrared region) the dispersion effects and the material losses due to wave propagation tend to cancel each other and create "windows" (meaning a really good transmission of those frequencies because the loss in the optical fiber has been reduced). Some experiments tend to show that going even higher in frequency than infrared may result in even better transmission characteristics and so improve the quality of the transmission over even longer distances than are now being accomplished.

The word "window" related to transmission is not a new concept. In submarine systems much experimentation has gone forth to find out what audio frequencies (SONAR) can best be transmitted through the sea water with the least losses. Some frequencies are much better than others, they permit much longer ranging with less power expended by the transducers energizing devices). The frequency must be kept relatively constant within that band that represents a "window."

As an aside, you are doubtlessly familiar with the concept of a rocket-launching "window." This means a particular time when the planets, the sun, and moon and such are all in a proper position to get a space vehicle into space for rendezvous with some far distant planet. Missing that window (or launch time) means that a long delay might result before an opportune time again arises for such a launch. What we want is the proper "window" for light rays to be presented to our I/O devices for the best communication, control, computation and such. This means that the output (O) devices must match the optical fiber for transmission of the proper light frequencies at the right *intensity* and with the *proper entry angle* and with the proper *polarization*. The (I) input device must be designed to accept that polarization efficiently to insure the most reliable, efficient, and cost-effective type of system possible.

OPTICAL FIBER CONNECTOR UNITS

As with any relatively new development, one finds various devices that can be used with the new developments that are not widely known. Perhaps the optical fiber connectors made by Seiko Instruments are such devices. Certainly we think about terminating the ends of optical fibers into something which may then be connected to an input or output device. We visualize that this connector must be strong and secure to provide the physical coupling necessary for proper light transference.

In Fig. 3-4 we show the generalized concept of this connector and we note the similarity to a regular coaxial type of connector for rf coaxial line of the flexible type. But there must be a difference. That is, the connector of Fig. 3-4 must provide a flat pressure connection to the transparent window of the sections (even though a "pin" type insert is used) so that the light rays can easily pass through this fitting. Seiko claims that this is accomplished in this type of fitting.

The unit consists of a plug ferrule, receptacles, and a fiber polisher for single-mode and multimode signal transmission applications. The input strand of fiber is the PVC covered type of 3 mm diameter and is a single fiber only. It is the model SAP-2 which is said to have a connection loss of only about 0.5 plus or minus 0.2 dB when in the single-mode transmission state using a 10 micrometer core optical fiber operating in the 1.3 micrometer wavelength region. In addition to the SAF-2 plug, a high precision ceramic plug ferrule Model SF-1 is required.

If you are considering some experimentation with optical fibers and would like to join several lengths together to cover a given distance, then perhaps the plug and ferrule arrangement shown in Fig. 3-4 will be of value. At least you then have some flexibility in

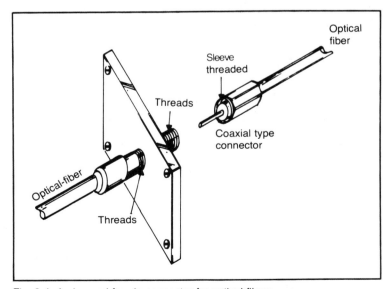

Fig. 3-4. A plug-and-ferrule connector for optical fibers.

assembling your optical fiber lines and although you may find some modifications necessary in order to accommodate fibers of other than the stated dimensions (so you can experiment with other than the 1.3 micrometer wavelengths) this can probably be accomplished with some advice and assistance from Seiko Instruments or an equivalent company such as Oriel Corporation.

THE OPTO-ISOLATOR AS A PART OF THE I/O DEVICE

We can probably consider the opto-isolator as an output device by referring to Fig. 3-5. Here we see how an LED or laser as part of an integrated circuit package is connected to some device such as a communications system, a computer, or whatever, which normally has an electrical output that can be adjusted to the proper level of voltage to operate the LED or laser, which produces a light intensity proportional to the signal level of the IC's input. This can be an off-on type of signal (binary type as from computers) to a modulation type (varying intensity in amplitude) as from a voice or other type communications set. The light reproduces the variation in that input electrical signal.

We must have a connector attached to the IC to accommodate some type of connection to the optical fiber line we will use. A plug and ferrule type connector such as just shown might be ideal. Next, we must have some kind of optical fiber line which has the proper core type (size and index of refraction) to accommodate the mode of transmission we desire, single mode or multi-mode, and accommodate the frequency of light we obtain from the output device shown. It's not just as simple as finding any kind of optical fiber and gluing the ends together and connecting it to a light source and a photodiode, is it?

The far end of the "line" or fiber is connected through another connector to the input device that must be some kind of "light to electricity" conversion unit. It might be an SCR if we are going to control something. It might be a photodiode if we just want a linear voltage output as we would want from a modulated input. It might just be a kind of photo-transistor if all we need is a signal when light is present and no signal when light is not present. We've listed some types of elements. There are others that your radio-parts dealer can show you or you can find others by examining catalogs from the various manufacturers.

In Fig. 3-5 we have shown what could be a connection from a computer to some device or from some device to a computer. With the increasing use of computer controlled robots (see TAB book

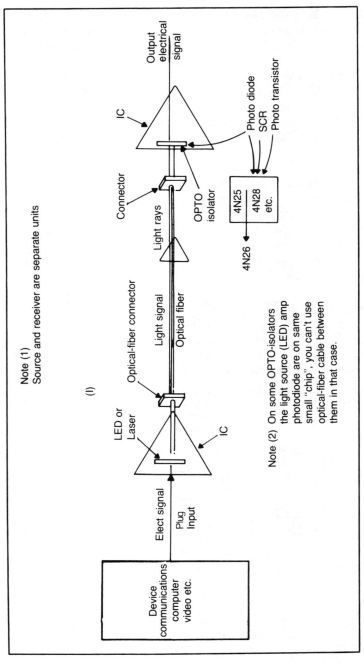

Note (1)
Source and receiver are separate units

LED or Laser

Optical-fiber connector

Light signal
Optical fiber

Elect signal

Plug Input

Device
communications
computer
video etc.

(I)

IC

Light rays

Connector

OPTO isolator

IC

Output
electrical
signal

Photo diode
SCR
Photo transistor

4N25
4N28
etc.

4N26

Note (2) On some OPTO-isolators
the light source (LED) amp
photodiode are on same
small "chip", you can't use
optical-fiber cable between
them in that case.

Fig. 3-5. Using opto-isolators as output devices.

85

numbers 1071 and 1421) and numerical controlled machines such connections between and among the various parts of the system seem very likely. Recall that the reduction of induced electrical interference is so great using optical fibers in the connecting-communications and data-transmission roles that this seems to be likely to be the next step in computer system accomplishments.

We need to point out that most opto-isolators currently de-signed are devices to accept an *electrical* signal input and provide an *electrical* signal as output. Figure 3-6 shows one type connection for these units. With fiberoptics we imagine that the interior of this device is "split apart" so that the light source and light receiver are the elements of the I/O devices with an optical fiber connecting them together as we have shown in Fig. 3-5. You can easily obtain opto-isolators from almost any radio-parts store and they have great value in actually isolating the electrical circuits of an output from the electrical circuits of another section of the device. They permit different levels of voltages to act in and operate within each electri-cal circuit independently of one another.

USING THE INVISIBLE LIGHT BEAM UNGUIDED BY OPTICAL FIBERS

We don't want to consider optical fibers so much that we forget that a laser ray may be propagated in a pencil thin line into space and be reflected back to some source. No optical fibers are necessarily used in this accomplishment. We know this is used in some survey-

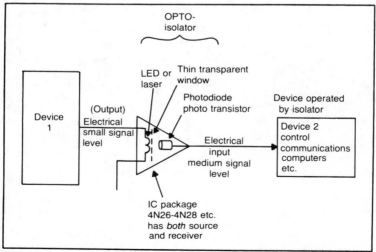

Fig. 3-6. A conventional opto-isolator circuit.

ing and leveling instruments and in ranging systems for the military and in some guided weapons. It is of some interest to us to know that at least one company, in close proximity to the missile range at White Sands Proving grounds, New Mexico (Energy Optics of Las Cruces) has developed a ray gun that uses an invisible beam of light to read utility meters from several hundred feet from the meter. This should really be no surprise. In supermarkets there are countless "bar readers" used by checkers to record the price and type of product being bought. They run the bar graphs over a special grid which uses light reflections from a small laser to an optical-electrical conversion unit to produce electrical pulses of wide or narrow width and spacing. These can then be interpreted by microprocessors as binary signals and thus numbers that can be converted into electrical voltages to cause various LEDs to glow brightly on the cost display screen on the counter. The device records what was sold, date of sale, cost of item, quantity of item, and stores it for print-outs for use by management. So, let us now examine the concept of one of these "bar readers" and find out how they work.

HOW A LIGHT READS A BAR GRAPH

Let us consider what we would require if we were designing such a system. We know we must be able to "read" the bar graph. Figure 3-7 shows such a graph and the basics of our reading system. Because the graph may take up considerable space it must be compressed and that means the lines must be narrow and the spacings between them very small. This means we need a light beam small enough to be able to distinguish between the thin lines and narrow spaces accurately. A laser beam is the answer here.

Second, we must be able to "flash-over" all elements of the bar graph in the small time it appears within the viewing range of the laser beam. Two ideas come to mind. We might have the person "drag" the package over the window and thus provide the motion necessary to move the bar graph past a fixed beam position. We might have the beam itself move and *scan* the window in some split second of time. In the second case it really doesn't matter whether the package moves or not as the terribly fast moving beam will scan so fast that to it the package seems standing still even though it is being moved by human hands.

Next, we must consider that the beam of light will be reflected from the bar graph on the package and must hit a receiver photodiode so that an electrical signal can be generated by that reflection.

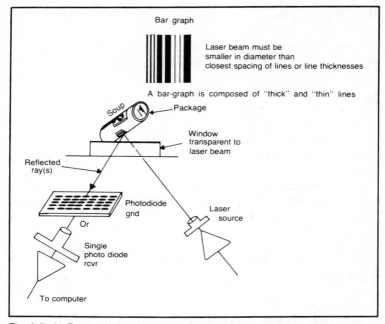

Fig. 3-7. A "Bar-graph" and a means for reading it using light rays.

When the beam passes over the marked line of the graph the reflection will be of low intensity thus the electrical signal will be small. When the beam "hits" the spacing between dark or black lines the light reflection will be strong and a large electrical signal will result. Now, all we have to do is to devise some way to insure that the reflection will always "hit" the photodiode (or photodiodes if we use a bank of them) even though the beam is moving. One system uses a revolving mirror system to scan and by its own geometry this mirror system presents the beam to the bar graph at such angles that even though it "hits" at different physical lengths along the graph, the reflections all come back to a given fixed location (see Fig. 3-8).

Next, we want to send the electrical output of the photodiodes to a computer to make a number that identifies the package as to size, content, cost, and such. The computer will then memorize the fact that one such (or more) packages have gone over the window and thus have been sold, and it will keep this data for reordering and inventory. It will, at the same time, perhaps, select the cost and display that for the purchasers anguish! So we need to convert the electrical signals into binary numbers. Various types of coding will

88

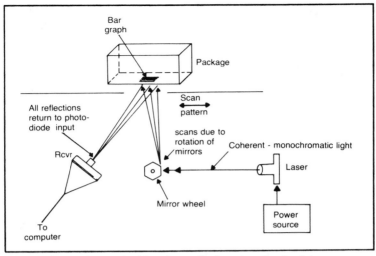

Fig. 3-8. A scanning method to return reflections to a fixed point.

make this possible. Some methods are shown in Fig. 3-9. Now we can design our own system and probably make it work. We know others have done so because these devices are in use currently in many stores of various kinds. A further note: In some applications

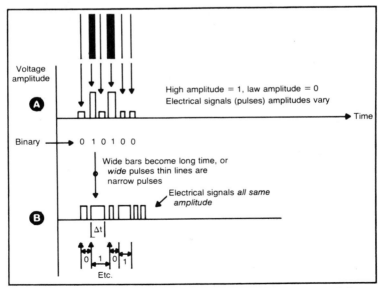

Fig. 3-9. A concept of coding of light reflections to get binary representations.

the machine is designed to give an alarm signal (not a "beep" but a deep sounding "boo") if the package graph has not been read for some reason. It could be because the graph was not flatly presented to the window and therefore could not be scanned, it could have been a fold-over on the package obscuring the graph—and, don't you know it—*the computer knows* when it has not scanned a full bar graph!

How does this fit into fiberoptics? Well, as suggested previously, reading a utility meter from a distance using a laser beam now is a reality (assuming the meter has a kind of graph made up of all its dials that can be read in this fashion) and that instrument can use fiberoptics to convey the light beam into the "gun" that may be pointed toward the meter. Another section of fiberoptics can be used to return the flash of reflected light to a detector deep inside some complex equipment carried on a shoulder or in a car or such. So, fiberoptics can well be involved in this new development and in countless other developments in the future.

Going back to Fig. 3-9 we need to consider some aspects of this illustration. First, in A we notice that the *amplitude* of the voltage output is what determines whether the computer will recognize a 1 or a 0 from the pulse. It works all right, but this system is very subject to interference! If a spurious signal gets into the system it could easily destroy the voltage levels and thus cause great confusion to the computer. It could make a 0 appear like a 1 by increasing the voltage output pulse level, and so on.

In B of Fig. 3-9 we see how this might be overcome. By keeping the pulses at the same amplitude (limiting them) the effect of interference is greatly reduced. Then, if the magnitude of the voltage or length of time the pulse exists is "stretched" slightly we can make use of a pulse width (duration) code that has constant spacing and can be very accurate. Other codes that might be used are pulse-presence-and-omission, pulse-position (in time) with respect to a fixed reference pulse for each pulse train, or graph reading, and perhaps even other methods that would require a different type of bar graph. The field is open, let your imagination run wild!

APPLICATIONS OF FIBEROPTICS TO ROBOTICS

Because robots may use fiberoptics in their extremities and for seeing applications, the use of optical fibers to convey the light reflections back to some central location where integrated circuitry could convey the light signals and intensities into usable electrical signals seems almost a mandatory requirement. We have mentioned the use of optical fibers in a robot's hand and we imagine them running to various locations in the "fingers" or "grippers" in order to locate objects that the robot is to take hold of, move, pick up, or whatever.

I have been informed by Sumitomo Electric Industries of Japan (with due regard for promotional flattering for their inventions) that they have developed a "most human robot" with perceptions of hearing, speaking, learning, and physically, with arms and legs to approximate, most diligently, the human torso. This "humanized robot" or as we think of them, the approach to an android, is said to be able to understand human voice commands to such an extent that you can tell it to go and pick up something and carry it somewhere else and place it there, and it will obey your command. It is said to be a small unit (shades of R2D2!) being only some 36 inches by 20 inches wide by some 39 inches deep. Almost as fast as most humans!

The information that seems most pertinent to us at the moment is the fact that a description (somewhat limited) of the robot tells us that it has two eyes that are moveable and that they are made of some 300,000 optical fibers. This permits the robot to distinguish "through image recognition technology" (see TAB book number 1421) the shapes and sizes of various objects, perhaps even the colors. It is also stated in the announcement that the robot has some optical character-reader capability in its legs that permit it to read written messages and instructions and to follow various paths as it moves and to permit it to avoid obstacles.

The state of the robotics art is advancing. But, you knew that didn't you? What you may not have realized is that the "eyes" of this robot (if everything stated is anywhere near factual) means a kind of "breakthrough" in object recognition by a machine of this type. That optical fibers are used to permit, first of all, a grid of light "pixels" (necessary to give the robot some vision capability) and that this grid might have as many as 300,000 elements, is quite an accomplishment! Second, the use of optical fiber strands to connect from the *moving* "eyeball" or light scanning unit, as it might be more properly described, to a fixed element which would be some kind of microprocessor with untold numbers of electronic elements such as resistors, capacitors, coils, filters, transistors, and other currently desirable elements to process the various light reflections and identify the patterns as objects, places and spaces, does seem somewhat marvelous.

We reflect on the control problem. Imagine what it might be like to design a servo-system to move such a scanner. Something small and some device that would use some of the reflected rays to govern just where the "eyeballs" would "point" or focus. Then too there is the concept of lenses, perhaps a single lens for the whole grid or possibly one for each fiber (and recall that these are "hair-thin" in size and each might have its own lens "built-on" to its seeing end). Also consider the multitude of micro-small connectors necessary to connect the distant ends of the fibers to the microprocessor unit!

It is not hard to imagine that a fiber - optic cable might contain a multitude of fibers and be relatively small in size, flexible, and quite easy to handy physically. The "scanner eyeball" would not have to be big physically. But, consider that this unit (like some others attached to a robot) might have to be moved anywhere within at least a hemisphere and you've got a control design problem of no small magnitude.

So, here is another use of optical fibers in addition to the long telephone lines now being installed by the telephone companies who use optical fibers for communication, data transmission, control and so on. This field of automation, robotics, and numerical machines is expanding at "break-neck" speed in the design of "seeing" devices for these machines so that they can be given instructions that we could follow. "*See* that tool over there? Get it, and fasten this bolt. *See* it here?"

We begin to realize that the day of wire connections between the various electronic solid-state devices may be approaching its end. We may now see in addition to printed circuitry type boards new boards that have fiber-optic cables and lines built-in to convey this new method of sending data and signals throughout the device. The use of metallic "electron roads" in electronics may well begin to reduce and even approach the vanishing point. So, we need to learn and experiment in order that when the time comes for us to confront this new technology we won't be caught napping. To return to our subject of robotic eyes let us examine Fig. 3-10 in which we show how the eyes of a robot might be fabricated. In A we show a round eye and in B we see more machine-like eyes that are a flat segment in a Cardan suspension system driven by two small servo motors and their associated electronic systems. Each type of eye may or may not have lens arrangements in front of the multitude of fiber-optic ends which pick-up the light rays from the reflections and convey them to the microprocessor located somewhere in the body.

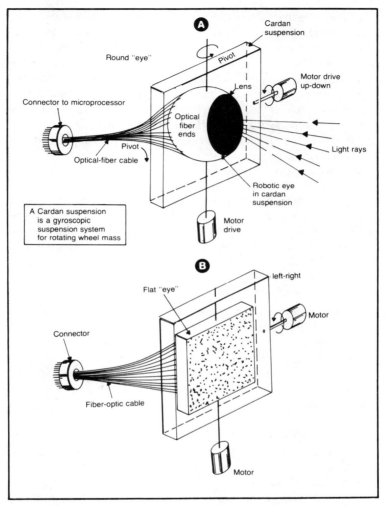

Fig. 3-10. Japanese robotic eyes using fiberoptics.

USING CAMERA DEVELOPMENTS IN FIBER-OPTIC APPLICATIONS

Moving the robotic eye so it can see things requires using servo-systems of extremely small size and great precision. We look to the field of photography for a solution to this problem. There we find a Honeywell development (the *Honeywell Visutronics Group* of Denver, Colorado). This is a small through-the-lens focusing system that is automatic and autonomous. It was designed for SLR cameras and uses two chips: an electronic eye and an electronic interface to its microcomputer. The light sensor chip is so designed that it produces two signals as output that are "in phase" electronically only when the camera lens is focused. Here's the clincher for our robotics application: This tiny microcomputer determines the distance and direction (in-or-out) to move the lens to bring the object into focus using very tiny servo-system of great precision. This system has an 0.05 millimeter focal distance and its accuracy equals human performance (so they claim) for high contrast scenes and is better than human performance with low-contrast scenes.

It is not without foundation to imagine such a robotic vision system as we have just described. All we have to add is *identity* of the light and dark objects seen, or the color geometry of the scenes and objects that the robot might look at, so it knows what it is seeing and then it can tell where they are and how they are oriented. There will be much use of robotic vision in forthcoming machines to make them more adaptable to our needs and desires. Fiberoptics will play a most important part in these developments.

MANUFACTURE OF OPTICAL FIBERS

If you want to make a copper wire you take an ingot of copper heat it and start "squeezing it down." As it begins to get longer and longer you keep it semi-round and finally you push a heated end through a small hole in a metal die and pull or "draw" it through that hole to make it come out the size you want it to be. Then it cools and you wind it on a reel and you have your wire. If you want it covered with insulation you run it through such a coating "bath" before you wind it up for use.

Making an optical fiber is done similarly. You heat some sand, silica, and perhaps a few other chemicals until they are molten. You stir them to get a good even mixture. Then you start forming the glass rod much like you formed the copper wire. In the "drawing" process with glass, you may have to add heat to keep the material the right plasticity so it will "draw properly" into those hair-sized optical fibers.

The various types of optical fibers are drawn out on machines that are all basically similar. Some very refractory type materials are drawn from solid rods or (as they call them—*preforms*) but the softer glass structures that are fluid at temperatures of only about 1200 degrees centigrade might be drawn directly from special containers or crucibles. What happens is that the glass is put into the machine which feeds it to a heating source and then after the glass is softened by the heat and becomes plastic it is pulled down through the "sizing die" to give it the proper diameter. Then it goes on down to a kind of wind-up drum where it can be stored neatly and efficiently. Recall that glass loses heat very quickly.

If you have ever watched a glass-blower at work and marveled at his skill you remember how he used the heat of his blast furnace to melt the silica, sand, and chemicals into a mass of molten glass. The color of the glass is determined by the chemicals and the type of glass-sand used. The glass-blower often stops to reheat the glass and then with quick decisive chops of a cutting tool he lops off unwanted segments as the glass object forms under his expert blowing and shaping. He can add to the glass by fusing new hot glass sections to the base structure and presently he has a vase, a swan, a bull, or whatever, which is at once delicate, amazing, and beautiful to behold.

So it is with our optical fiber drawing machine. It simply adds heat at the "stretching place" so the glass fiber will become plastic and can be drawn out into its tiny diameter. Some machines monitor the fiber size to insure that just the right diameter is maintained. Also, as with the wire, the coating of the core structure may be accomplished as the fiber (correctly sized) is drawn down toward its "roll-up" drum. Coating helps to preserve the fiber's tensile strength.

The glass used in optical fibers is said to be softened at some 800 to 1200 degrees centigrade. This range of temperatures must be provided to enable the drawing of the fiber into what we would consider to be a usable form. In a method called the "double crucible" or "double-bushing" method of drawing the fibers, the core glass is placed in an inner crucible which has a hole in its bottom and the cladding type glass or material is placed in the outer crucible. It also has a hole which is concentric with the inner crucible's hole. As you'd imagine, when the inner glass starts to flow, the cladding glass also starts to flow and it "wraps around" and "fuses-to" the core glass. The cladding material has a different index of refraction from the core and is made to become a part of the optical fiber strand. Using this method with various types of physical arrangements cores with graded indices of refraction can be obtained. This can result in single-mode transmission and low dispersion loss of the light rays through the glass material. It is said that the two crucibles are fed with preformed rods of glass which have the required chemical characteristics for proper indices of refraction as desired by the user. It is an automatic process once the machines are started and placed into operation. The cladded glass fiber is finally wound on a storage drum for shipment and use.

There are some types of optical fibers that are drawn from specially prepared solid-glass rods. Temperatures over 2000 degrees C. are required to soften the high silica-content material to a plastic enough state so it can be drawn into the small diameter fiber. This high temperature requirement may require special heating devices. Induction heating furnaces are one type that have been successfully used to obtain the required temperatures. Also, when directly heating the rods, special lasers of the CO_2 type have been used. It is said that using lasers gives a fast temperature response (quick heating) and the ability to draw out the fibers in almost any atmosphere results.

There are multi-component materials used to make fibers. These may be the soda-lime silicate and sodium borosilicate glasses. These are prepared by melting silica (that makes the glass) and adding some of the other modifying compounds to the *melt*. The additives may be carbonates or nitrates which decompose to form the oxides in the "melt." It is very important that only very small concentrations of impurities be present in the resultant optical fiber if it is to have a low-loss light transmission capability. This has to be true especially in that region of color or wavelength that is to be sent over the fiber. To insure an adequate purity of the fiber, the materials themselves must be "super pure." This requires sampling during preparation to insure that only the purest kind of materials are used.

It goes without saying that in addition to having pure compounds to make optical fibers, the methods used to melt the glass must also insure that the fiber will be of adequate quality. This means it must be chemically and optically homogenous and free from bubbles and inclusions. An inclusion is a fracture in the glass structure. To get the required homogenity (mixing of the chemicals evenly and thoroughly) the melt is stirred or a stream of gas is passed through the melt during the melting phase of manufacture.

In one example of the manufacture of an optical fiber, a pre-mixed silica, sodium carbonate, and calcium carbonate of high purity was melted down in platinum crucibles using rf power (like our microwave ovens) to heat the crucible. After the substances were melted, they were stirred with a platinum paddle at 1500 degrees centigrade. The melt was then permitted to "soak" at this temperature (or simmer) until large bubbles floated to the surface and burst. Finally, more stirring was done with the special paddles until the required homogeneity and optical purity was achieved. The temperature was then reduced some 200 degrees centigrade and again a "soaking" time was used to permit ultra-small bubbles to be absorbed into the melt. The whole mix was permitted to cool down to about room temperature and annealed (heated and cooled) over a time period of some 20 to 24 hours. If there are high losses in light transmission, it is said to be because contaminations get into the material or the material is not optically pure. Thus, only the purest materials are used and the utmost care in the manufacture of optical fibers is taken at all times.

It is of interest to us that rf power is used to melt the materials used in optical fibers. This leads to fusing of the ends of fibers themselves so that a continuous line can be made that will convey the thousands of conversations, etc., which companies like the *Bell Telephone* are interested in developing. The electrical conductivity of the glass used when it is in the molten state, is high enough so the rf power can couple directly into the molten glass material and heat it. Another interesting aspect of manufacture is that if the melt is cooled quickly by water or similar means after mixing is completed, then the melt does not tend to absorb impurities from the crucible walls and thus the melt tends to have the same purity standard as the materials from which it was prepared. It is said that some types of fibers have losses of 27 to 50 dB/km when prepared and these losses are at a specified frequency of around 1080 nm. There are many problems in getting the melt to be consistent and smooth and homogeneous. One problem is that if the melt is heated a little more in one location in the crucible than in other locations, convection currents tend to set up in the melt and this makes for bubbles which are not easily detected and removed. It isn't just the simple process of "heat and shape" as the glass blowers employ. Here the heating must be uniform, the right temperature, the materials of the right purity, and the melt must not be subject (any more than it is possible to prevent) to outside impurities getting into the melt as the fiber glass is formed into the strands that we will use for all our purposes.

One of the materials used to make optical fibers is sodium borosilicate glass doped with thallium oxide. This has been called the *Selfoc* type of fiber. The cladding of this type fiber is formed from an undoped sodium borosilicate. The technique used is such that as the fibers are drawn (double crucile method, so that the cladding is over the core material) the cladding material is in contact with the core when it is very hot for some period of time. What this does is to permit an exchange of ions between the sodium and the thallium and when this happens a *graded refractive index is formed*. This, we know, enables one to have a monomode type of transmission of light, so having a graded index of refraction can be of great importance. The losses in transmission using this type of fiber have ranged from 14 dB/m to about 20 dB/m when the light frequency was in the 800 nm range.

Another material used in the development of optical fibers has been a synthetic material that really is a fused silica. It seems to have permitted good passage and low attenuation of light in and near the infrared region. The synthetic fibers were prepared from synthetic silica made by a process of vaporizing silicon tetrachloride and subsequent oxidation or hydrolysis. The material produced was said to have a very high purity. There is always a "catch". Because very high temperatures are involved, a special technique called *"soot deposition"* using a *white soot* that is formed when gaseous silicon compounds are decomposed in the presence of water is used. Using this soot in a special process that first caused the soot to be impregnated on the inside of a glass tube, and then drawing the tube into a solid fiber rod resulting in the production of very efficient, low loss, fiber-optical light guides.

Still another method of manufacture has been used to develop pure and low loss optical-fiber light conductors. This has been called the *chemical vapor deposition method*. In this method a layer or many layers of special material is deposited on the inside wall of a tiny glass tube. Then the tube is heated until it becomes plastic enough to collapse on itself and it becomes a rod instead of a hollow glass tube. All that remains is to "draw" the rod through the proper die-type material or former until it reduces to the desired size and diameter. In this method fused quartz has been used in the deposition process. It was found that in this process the core material had to be further doped with some oxides so that it would have a *higher* refractive index than its cladding glass, so this was done using some fused silica that was doped with other oxides. The core might have been made originally from germanium oxide doped silica. Later, other materials have found consideration as a core material. It is interesting to consider a chart that shows to some extent a "window" and some attenuation characteristics of some glass that has been made from many materials (see Fig. 3-11).

The problem of pulse dispersion in an optical fiber has been discussed in earlier pages. Essentially it means that for some kinds of fibers and for some propagation methods into fibers, the output

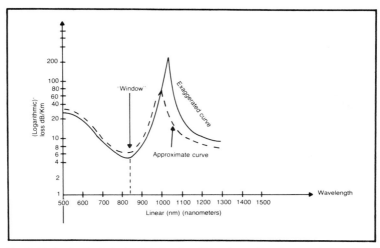

Fig. 3-11. An "optical window" defined and the attenuation characteristics of one type of light-fiber guide.

pulses are not distinguishable (in the worst cases) and are badly distorted in other cases. This is due to the fact that not all of the light frequencies involved in the pulses are transmitted in the same manner, with the same efficiency, and with the same phase or timing. So when recombined in the fiber's output, the signal can actually be a "mess" instead of nice, clear, sharp pulses. We know that optical fibers in which the refractive index changes gradually (from the cladding toward the center of the core material in a precisely pre-determined manner) will have less pulse dispersion than a nonvariable refractive-index type of fiber. In manufacture it is desirable to be able to vary the refractive index in accord with the scientist's specifications to accomplish the best possible transmission of light. There are machines that can accomplish this graded refractive index requirement by using a deposition method and many layers of various types of chemical materials. There are several types of flourinated polymers that have lower refractive indices than the fused silica and they can be used as cladding type materials. It is said that they may be "lossy" and we are not interested in losses—quite the contrary!

In the manufacturing process one looks at the size of the fibers. Much work has been undertaken to develop monomode types of fibers that have good tensile strength and are large enough in diameter to facilitate splicing. If a telephone company wants to run lines that are miles in length, then they want very low losses, good transmission characteristics, and the ability to easily splice the optical-fiber cables and lines. Make a note that it is not easy to develop an optical fiber which is mono-mode at many light frequencies. In fact, it may not be possible. But some bandwidth is permitted in this type mode, such as a band around 1060 nanometers wavelength. Within this band the communications capability is almost beyond imagination.

THE BANDWIDTH CAPABILITY OF LIGHT CHANNELS

Let us do some analytical work for a moment. We have heard about the tremendous communications capabilities of the extremely high frequencies such as involved with light channels. But we have not examined this. Let us assume that we have a wavelength of 1060 nanometers, meaning 1060×10^{-9} meter. It can be written 1.060×10^{-6} meters or a little over one micrometer. To find the frequency we divide this into the speed of light (c) which is 300×10^6 meters/seconds.

$$f = \frac{c}{\lambda} \qquad \lambda = \text{wavelength in meters}$$

$$\text{therefore } (\therefore) f = \frac{3 \times 10^8}{1.060 \times 10^{-6}} = \frac{3 \times 10^{14}}{1.060}$$

or, for our purposes, we might approximate this as something a little less than 3×10^{14} hertz/second. That is a big number. If you don't think so, write it out, use a (3) and 14 zeros after it!

Consider that a light frequency *band* might include a few frequencies each side of this center frequency, or say that the *band* is from 3×10^{14} to 3.5×10^{14} hertz/sec. The factor we want to know is; "how many channels of, say, 5,000 hertz plus or minus some 3 kilohertz will "fit" into that higher frequency band of 0.5×10^{14} hertz/second. A "rough" mental division will produce at least 0.16×10^{14} channels of the audio-spread bandwidth. That's a lot of channels! And that is only the beginning! If we use several light channels, each with the previously indicated capability, we've got enough to handle *almost* (some weasel-wording) anything now or in the immediate future and that includes TV and other types of video signals and data signals for computers and control signals and such. Yes, there is a decided advantage to the use of light frequencies as channels.

Some will undoubtedly question the narrow channeling we

have used as an example for the audio channel—the 5000 hertz. Some, and perhaps one are you, might think that a much higher range of frequencies might be needed for, say, pulse-type transmissions. You are right, of course. But we ask your indulgence, for if you know that, then you also know the capability of light channels perhaps better than most and do not question that capability. For your own satisfaction you might recalculate how many channels you believe are available on a single light channel. We'd like to know your answers.

CONSIDERING THE MANUFACTURE
OF OPTICAL FIBERS AND THEIR STRENGTH

You have some ideas as to how optical fibers are manufactured and how they are made with a single index of refraction core and how they might be made so that a variable index of refraction is obtained in the core and cladding. Now we want to consider the very practical aspect of manufacture relating to how the fibers are made so they have sufficient strength to be "drawn" out and still stay all together when they are only a few "hair diameters" in thickness.

The problem with the manufacture that affects the strength of the fiber is that there are flaws in the manufactured glass of various types and these limit the strength of the fibers as to stress and strain. The closer a flaw is to the fiber surface the more it affects the total strain capability. The literature one might study in connection with the failure of glass structures varies considerably as to the individual testing results. This could be because there is a difference from batch to batch on the fiber material or because flaws creep into one batch and may not do so to the same extent in other batches or bundles of produced fibers. It has been noted that even if a glass fiber does not break when in production it may break later. Surface cracks of a size smaller than the critical size for a brittle fracture may actually get larger under some conditions (for example if water vapor is present) and they grow until the fiber actually breaks. This kind of fracture is sometimes called a "fatigue" factor and it may take place after many years of existence of the fiber.

We can also imagine that the strength of an optical fiber might be related to its length and we wouldn't be wrong. It is just natural that the longer the fiber produced, the more likely it is that there will be some flaw generated or appearing due to "circumstances beyond production control" that will later cause fracturing problems. This is also true with wires that are drawn. If you take a small segment of either one and tie it between two terminals, it might be okay. But take a very long segment and tie it up between terminals or put any strain on it in any way even if it is lying in an earth-hidden cable and sooner or later it will develop troubles.

Every effort is made in manufacture to make fibers strong. Every aspect of the manufacture is carefully controlled. In the handling and winding of the fibers care is taken to prevent abrasions which might "nick" or "scar" or otherwise affect the optical fiber surface. Sometimes the fibers are coated with some material that will add strength and still will not affect the light ray transmission capability of the fibers. These materials are coated over the cladding. But, the literature tells us, the most damaging element to the fibers is water vapor which, they say, produces fatigue or failure in some reasonably long-term time span. When fibers fail or have to be rejoined the splicing processes we have previously mentioned will take place. These may be done by machines, or men, or both. The methods include fusing the ends of the fibers together, "gluing" them together with special compounds after making the ends perfectly "flat", or using connector devices that pressure-hold the ends together. In any event the method used will be that one that gives the best strength and the least loss of light ray transmission.

It is said that clad-glass fibers do best when they are "microscopically" straight. They have the least light loss that way. There are no bends in the fiber to start up dispersion and multi-mode effects which reduce the effectiveness of the light transmission. You wouldn't believe that (as some literature describes) if you wind a fiber on a drum surface *that surface may not be perfectly flat* and so the imperfections on the drum surface (that might just be dust particles, little metal chips of microscopic size, or just scratches on the drum surface) can reduce the light transmission capability and produce losses. But, if the fiber is coated with some kind of protective coating before it is wound, these losses are minimized to some extent. The problem, recall, is that the optical fibers are so small in diameter that *almost anything* can bend them in some microscopic manner causing the loss problem. Some companies such as *Northern Telecom* have designed a special cable arrangement that is (looking at it from the end) a segmented, multiple clover-leaf pattern where each "leaf" is a slot. Into each slot is placed the optical fiber, or fibers, in such a manner that they are not under the stress and strain of the cable material itself. Nor are they subject to the microscopically uneven surfaces that might be present against an unprotected optical fiber. In some of the slots are placed wires for those control operations which demand such connections, so a "cable" is a rather larger (but not too large) type of hose device, looking at it unopened. Inside are the "works" as protected as they can possibly be against all the elements and events which might cause problems in the light ray transmission over the interal fiber light carriers.

Some of the literature, however, suggests that rather than bunching the fibers in cables, these should be put into small groups which then would be formed into cables. The two types of cables suggested are the flat type (you are familiar with this "ribbon" type wire connector cable used with computers) or into the more conventional circular cable type. In either type the fibers, which are specially coated, are wound around a special central coated steel wire for strength, in a helical manner. The whole thing has a layer of polymer protecting the interior and the ends are encapsulated in urethane and finally covered with something like polyvinyl-chloride plastic. In the ribbon structure the fibers have to be separated enough to permit splicing and the ribbon material in which the fibers are embedded must be strong and protective like the steel-core plastic-covered circular cable.

SOME NOTES ON SPLICING OPTICAL FIBERS

When the ends of two optical fibers are to be spliced, the first consideration is to do this in such a manner that excessive losses of the light energy will not take place at the splice. This means for one thing, that the ends of the fibers must be as nearly "flat", which means they must be cut perpendicular to the fiber length, as near perfectly as possible. Consider that the fiber is a type of glass and imagine how you would make this "cut."

When you cut a piece of glass you use a scoring tool which scratches the surface of the glass. Then you tap the glass somewhat sharply near and on one side of the scoring line while holding the other side flat on a cloth covered table. It is always an amazement to us to watch an expert glass-cutter at work. They break the glass so quickly and cleanly. We've tried it and the result is almost always a jagged fracture which may, or may not, follow our scored line at all! We've wondered whether we made the scoring line deep enough, or too deep or what? Oh well, each to his own trade!

So we were not surprised to find that scoring the ends of optical fibers to cause the glass rods to break cleanly and in a flat, perpendicular manner was not always successful. Then we learned that a special machine was devised to do the job and this machine might use a diamond or carbide type of tip on the scoring tool to make a good clean mark. We found that the manufacturers were able to do this cutting quite effectively but the preparation of multiple fiber ends was another matter. Cutting them in the required "flat" manner, then polishing them and/or grinding them down to a perfect fit was quite a task. You can imagine that with a computer type cable which might have many optical fibers in it, the joining of the ends to connectors (or terminals) or splicing them onto further cable "runs" might be a very necessary job to do. The problems could be large in this respect.

Earlier in this text we have suggested a splicing liquid which might be used to "glue" the ends of some fibers together. This also might require some "doing" because you have to align the ends of the fibers which are to be glued, and then compress them together tightly until the glue "sets", and then perhaps cover the joint with some kind of tension protective jacket. It is said that if you can align the fibers and properly glue them together with an index of refraction-matching liquid glue, then you might not have losses over some 0.045 dB for some types of fibers, notably the high silica multimode low-loss types.

We have mentioned some types of mechanical connectors and these, may be either circular or "V" types and they simply "butt-join" the fibers together and a pressure fitting is tightened so they continue to press the ends together so the light rays can pass through this type joint. Some say there is an advantage to this type of splice in that you can pull-out and push-in the fibers to make the connections any time you choose (like pushing a copper wire into a connector and pulling it out when you want to change the circuit arrangement). If you are not careful with your fibers to make a perfect butt-joint fit each time, then you have severe losses or perhaps no transmission of light rays through your connection at all.

The companies have developed another method. They take a bit of plastic and "groove" it so the fibers will lie in those grooves. Then they glue a top plastic piece, which matches, in place and then the joint is secure and solid and cover-protected. When this is done properly, a very good continuation of the light transmission capability is maintained. Otherwise you might have to have a magnifying glass or microscope to make sure the fibers are properly joined together. Remember that they are extremely small hair-like strands of special glass, even with their cladding and some plastic coat covering.

We learn that to make a splice requires that the fibers be prepared properly for the splicing operation. They must be aligned properly so they mate properly and they must be somehow connected together as closely with an index-matching glue, or physically or in some manner such as fusing with heat in such a way that the integrity of the fiber is maintained even though the glass is heated to a molten state. Notice that we have mentioned several times that the "index of refraction" must be maintained with the joining compound, if any is used. That is no small accomplishment!

HISTORICAL USES OF OPTICAL FIBERS

As far back as 1977 an installation in Chicago by the *Bell Telephone System* connected two central offices and a customer location that was in the Brunswick Building to demonstrate the capability of optical fibers in communications with voice, video, and computerized data signals. Lasers and LEDs and avalanche photodiodes were used to transmit the light rays and convert them back to electrically usable signals. The system was so successful that an effort to use optical fibers almost everywhere began at this time. Bell-Northern Research is doing a fantastic job in the development of this type of device in preparation for the jobs ahead in conveying intelligence from one place to another.

We are reminded of some concepts which make the reason for this "excitement" more understandable. It has been said that the Gutenberg Bible can be transmitted over wires completely in one hour. That same amount of material can be transmitted over an optical fiber system in one second. Word comes from the Bell Telephone laboratories that special optical fibers can now be produced as easily (and much better in quality) than ordinary glass fibers. These are called "multimode" types of fibers. It is said that the glass fibers are just some 49 microns in diameter (25,600 microns to an inch). Also it is said that the single-mode fibers are merely 5 microns in diameter and this is less than the diameter of a human hair. Pull a hair out of your head and examine it. You are looking at something about the size of an optical fiber. You'll then realize how difficult the problem may be to get light rays into the end of this fiber and to get the light out into a proper detector device. You'll begin to understand why we have mentioned microscopes when thinking of working with such fibers and some difficulties in splicing and handling them.

It is said that some 37,000 miles of optical fibers have been installed by the telephone companies and are now being used! It is said that these fibers are the multi-mode types and thus are slightly larger in diameter than the single-mode types. When you go from a few microns to several microns (i.e. from single-mode to multi-mode) it is not a big "size" jump for we humans. Once again we emphasize that just as you had to learn to "deal" with wiring and wire cables and then with printed circuits, now in our future is the next development in this category—the optical fibers of the future. We have to learn all we can about them and if possible get some "hands-on" experience with them.

It was planned that the first attempt to use light frequencies as data channels would be to use them with the so-called DS3 rate of digital communication which, in North America operates at about 44.8 megabits/second and is used by the telephone companies in their "trunk" cable connections. If the rates of transmission are lower, then copper cables do the job well. But the effort is made to develop optical fibers so they can be used with data busing. Bell Telephone Labs has demonstrated a capability to transmit as high as 420 million bits over a mono-mode optical fiber system which was over 50 miles in length, without repeaters or amplifiers being required for this distance. This tells us something about the perfection of splicing methods used in this long cable!

Some of the other problems associated with optical fibers, now that manufacture and splicing methods have been brought to a usable stage of perfection, are that the signaling techniques using binary procedures make it difficult to design lasers and LEDs and photodiodes of various types that can follow the extremely fast pulsing rates used (especially when transmitting the megabit rates). Not only do the light sources and the receivers have to be able to create and accept the pulses, they also have to be able to do it while retaining the pulse shapes, and differentiating among the pulse "trains" or groups that make up the various kinds of intelligence being transmitted. This is not a trivial task!

Bell-Northern Research, Ltd. has transmitted digital pulses at the DS1 rate (1.544 megabits/sec) over trunk lines as long ago as 1962. It is interesting that this conglomerate found that the following requirements have to be met by interfacing procedures and devices and techniques and this must be done electronically in order for optical fibers to actually be used in communications in the North American digital world.

Designation	Digital Rate	Channels VF Capacity
DS1	1.544 Mb/sec	24
DS1C	3.152 Mb/sec	48

Designation	Digital Rate	Channels VF Capacity
DS2	6.312 Mb/sec	96
DS3	44.736 Mb/sec	672
DS4	274.176 Mb/sec	4032

If the optical fiber systems can operate at these rates, then they can be interfaced with proper devices to the current telephonic communications systems. Of course, there is always the question of "cost-effectiveness" of the optical fibers rather than coaxial cables, copper wires, radio channels etc. That has to be determined and then the choice may be based on the cost effectiveness figures rather than on any technical or manufacturing or maintenance conditions or developments. Well, you can say that cost-effectiveness includes such things as these and you'd be right. But there are other considerations in a cost effectiveness study.

It is of interest to us here, also, that in laboratory experiments, wavelengths as low as 1500 nm have been used for transmission very successfully over monomode fibers. The transmissions have been over very good distances with low enough loss that repeaters were not required.

It is written in company reports that because repeaterless spans might be used over some distances—up to 12 kilometers—that using a 850 nm wavelength with laser light sources and APD detectors that 60 to 70 percent of the urban digital spans could be repeaterless. Optical fibers used might be only 25 mm in diameter and since they are very small and lightweight (even when properly packaged with cladding and plastic wrapping and in cables) they can be placed on reels which contain much longer spans than an equivalent copper wire reel would reach. The cost seems to be less than that of existing type cables and this becomes very attractive to company installations. Some optical fiber strands are just 50 microns in diameter across the core material, have a cladding of a lower index of refraction and then are covered with a plastic jacket which builds up the size to some 330 microns in diameter (outside measurement). With such a fiber, recall that the light rays must enter the core within the cone angle of some 20 degrees, (10 degree half angle) and this gives rise to what is called a *numerical aperture (NA)* which is the sine of that angle or 0.18. It is stated that the larger the NA the better the light will propagate through the fiber but this has to have a "trade-off." One must balance the losses and distortion that come about with a large NA with the increased efficiency of coupling and the better transmission capability. Proper wavelengths with good "windows" in the optical fibers being used improve this relationship and make the "trade-off" less important. The optical fiber light frequencies are in the range from 800 to 1800 nm and some really good windows have been produced at frequencies of 1300 and 1550 nm wavelengths. Losses through these windows are said to be less than 1 dB/km and that is excellent!

The use of optical fibers to connect computers is becoming more and more important. The Japanese (who do much experimentation and have done much development in automation, robotics and computation) now have under development a laser printer which, it is claimed, can print up to 21,200 lines per minute! Also, they are working hard on development of optical-fiber cables that can handle infrared frequencies over which computer data can be transmitted from one computer location to another. It is possible that a breakthrough in optical-fiber chemistry might occur which could mean even better light transmission with more "windows" and less losses in the future.

We refer again to a growing problem in data transmission systems that use light rays. That is the problem of having light sources "turn-on" and "turn-off fast enough to handle the bits of data so that they won't be lost, distorted, or mish-mashed together so much that the data cannot be transmitted or recovered. We look at Fig. 3-12 wherein is shown the *relative* operational characteristics of a laser light source versus an LED light source.

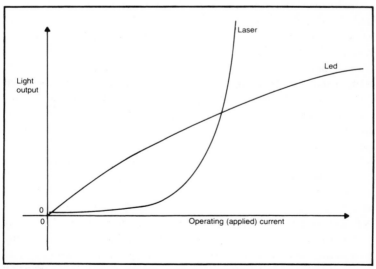

Fig. 3-12. Laser versus LED "turn-on" characteristics.

Notice that the LED characteristic curve approaches a some-what linear span which means that for a little increase in current applied, there is a very small increase in LED light output. Thus, to get a significant "burst" of light, representing a digital pulse, a really large change in applied current to the LED must take place.

On the other hand, notice that for the laser type light source, if you operate on the base of the exponential type curve, near the vertical ascent, then a small increase in electrical current can trigger a large burst of light energy and that means that in all probability a faster rate of pulse generation can take place. This assumes, of course, that the light source will "turn-off" as com-pletely and as quickly as the current can be made to decrease. So that if the applied current is a square wave then the light burst should also have a "square wave" type time duration in the optical fiber.

We go to the opposite end of the system and some types of systems which have been under development by Bell-Northern Research are shown in Fig. 3-13. At A we have the classic "one way" system. The energy goes in on the left and goes through the fiber to the receiver on the right. At B is a multiplexed system, which may be time-share or frequency-sharing in operation. As shown, the system would be frequency sharing as the input is modulated with the summation of input signals and on the output the signals are separated by appropriate filters. At C is the two-way system as used by many computer systems.

The fiber optical-connecting-link doesn't care which direction the light energy enters or leaves the fiber. It can transmit in each direction if appropriate electronics and connecting interfacing units are used to insure that it isn't trying to transmit both ways at the same time! Normally, a burst of pulse data would go from left to right then a burst of pulse data would come back from right to left. When this is done at an extremely high rate of speed (with nano-second timing in the trillionths of seconds) then the system, to all intents and purposes is said to be operating "two-way" continu-ously. As you can well realize if mechanical displays and electronic screens display the data then they are actually much slower than the transmission times so some storage system should be employed to "hold" the data until it is needed and used.

You might well think that if the optical fibers are so small, why not just use two of them on each link, one for transmission each way. That way you'd have the equivalent of separate lines for each direction of transmission and you could actually have simultaneous

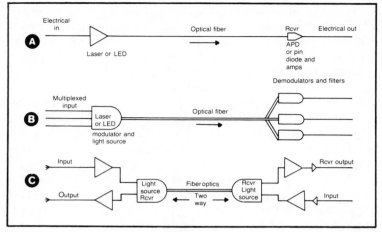

Fig. 3-13. Block diagrams of some fiber-optics systems (courtesy Bell-Northern Co.).

transmission going on all the time in both directions. Certainly this is a possibility, just add a link as at Fig. 3-13A and turn it around, and put two strands of optical fibers in a cable instead of one and the job is accomplished. Good thinking!

Now, in order to complete our thinking and imagination-generated ideas, we need to think about reducing the delays that come about in the *electronics* of the connecting and using units of the transmitted data. There are inherent delays in coils, filters, capacitors, and even in transistors and such solid-state devices. If we are ever to come near the speed of communication and computation of the human brain all these delays must be reduced and/or eliminated. Goodness! Isn't it strange to be thinking that the movement of electrons and holes in electric current transmission is *slow*! But it's a fact, and that is what the future of this science is all about!

COMPUTER PROBLEMS WITH OPTICAL FIBERS

Using optical fibers in computer applications (as of this date) has some problems according to studies and reports now circulating. One of the major problems is the fracturing of the fibers when they are used in cables and in connectors. This will be overcome by making a better product as time progresses and by improving the manufacturing techniques as related to computer and automation applications. There are some differences in how the optical fibers are used with communications equipment such as required by telephone, video, and long distance data transmission, and the close, short, intricate and delicate applications of a close-in computerized situation. But, research will come up a winner again as it has in the past, given the time to do so, and the money to invest and experiment.

Of course, all the associated equipment (light sources and receivers and interfacing equipment and such) that must be used with computers must also develop at a commensurate rate. But, one thing is certain: with all the advantages, operation-wise and cost-wise and efficiency-wise and time-span-wise, there is no doubt but what those micron-diametered optical fibers and all their associated connectors, splices, interfaces and monitoring and checking and operating equipment will come to be the prime area of consideration within computers now and in the future. In the next chapter we tackle something practical you can "get your hands on."

Chapter 4

Experimental Applications of Fiberoptics

Before we get too deeply involved in the theoretical aspects of light transmission, "light-guides", light generators, receivers, and such, let us take a breather for a moment and consider some uses and applications for light beams and light rays with which we might do some experimenting. We have not conducted the experiments, nor fabricated the equipment described herein. But most of what we describe has been built and fabricated and operated by some whom we know to be firsthand investigators. But first a word of caution. If a device doesn't do exactly what you'd expect it to do when you build it, do not hesitate to change and modify it as necessary to make the "end justify the means"! That is more than half the fun of experimentation. Being able to detect and become sensitive to small events and happenings and making them grow and expand (if this is desirable or eliminating them if they are a detriment) to enable the final product to do what you would like it to do. Keep a notebook. If you choose, let us hear from you about your efforts.

A LIGHT-FOLLOWING ROBOT

Both of these ideas are exciting. The fact that it is a robot (of one form or another) and that it uses our phenomena of light for some kind of guidance make it worthy of our consideration. We thank Karol and Mila Nowakowski for the information.

Karol and Mila have a robot with a phototransistor in its base structure. When the robot is in motion, the phototransistor senses an illuminated line on the floor. The base of their robot has a black felt skirt to conceal the sensors and other equipment, and to serve a more practical purpose, that of eliminating ambient light from the robot's surroundings.

The photo-transistor is set into a section of black (we assume rubber) tubing so that it has a narrow field of vision as shown in Fig. 4-1. We have shown in this figure a typical circuit that will produce a current output when it has a light input. If you are going to steer a robot in this manner you actually will need two such sensors. One sensor must be positioned so that if the base goes to the right, the electrical output of the photodiode will (through suitable amplifiers) be capable of operating some kind of relay or control circuit to cause the steering motor to turn the base so it moves to the left, or back "on track." The same effect (but in reverse direction) would apply to the second photodiode sensor. That way steering is accomplished. When both photodiodes either get the same amount of reflected light or no light, the base steering control unit should make the steering wheels "center," and the base thus moves straight ahead.

Back to Karol's and Mila's discussion of their robot. "The base of the robot has a small battery powered ultraviolet fluorescent light source. The guide line on the floor is made with invisible ultraviolet fluorescent ink of the same type used for readmission hand-stamps at various social and commercial functions. Thus the line is invisible under normal viewing. The robot can move around following a well defined path which we cannot see! How magical that seems to the uninformed! We suppose the real trick here is to get around the house when the lady-of-the-house is not home and paint our lines on the floors. If she saw that operation she might take a dim view of our scientific efforts!

Karol and Mila furnished the sketch shown in Fig. 4-2 which shows the base arrangement of their planned robot. They suggest the *FPT-100* as a sensor because it is much faster in response than a CdS cell. If the FPT is used, however, it is suggested that it should have a U.V. light filter over its face section for improved light differentiation. But, they say, a CdS cell might be used if the response time is not critical. The cell has a low sensitivity to U.V. light and a high sensitivity to the primary blue-green light *given off by the U.V. ink* of the most easily obtainable and common type. They sent us a curve showing a nice "window" effect for the CdS photocell around 5000 angstroms. You will want to recall that one Angstrom unit, whose symbol is Å, is equal to 10^{-8} centimeters. For example the green line in a mercury spectrum has a wavelength of

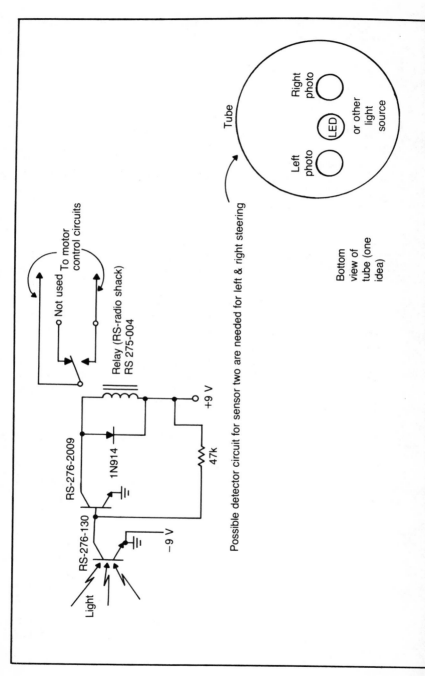

RS-276-2009

1N914

RS-276-130

Light

−9 V

+9 V

47k

Relay (RS-radio shack)
RS 275-004

Not used — To motor control circuits

Possible detector circuit for sensor two are needed for left & right steering

Tube

Left photo

LED

Right photo

or other light source

Bottom view of tube (one idea)

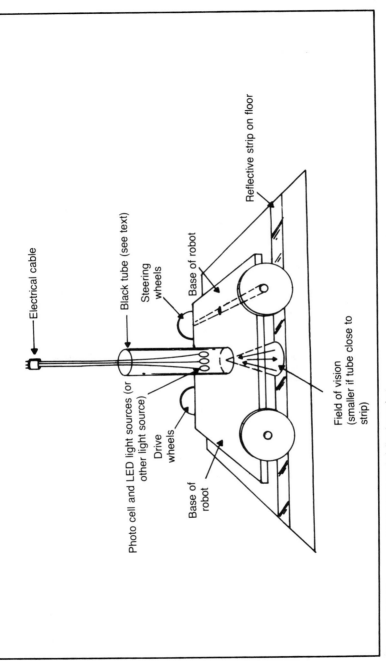

Fig. 4-1. A light sensing concept for a path-following robot.

Electrical cable

Black tube (see text)

Steering wheels

Base of robot

Reflective strip on floor

Photo cell and LED light sources (or other light source)

Drive wheels

Base of robot

Field of vision (smaller if tube close to strip)

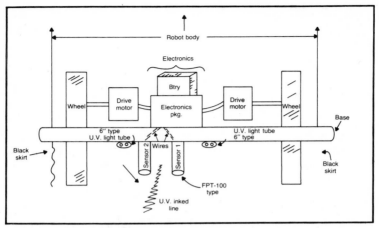

Fig. 4-2. The Nowakowski line-following robot idea.

0.00005461 centimeters, or 5461 Angstroms. We show the curve in Fig. 4-3.

By the way, we are informed that it is easy to change the path once you get the system working. All you have to do is to wash up the old line from the carpet or floor or whatever and "ink" down a new line. Also, if you have the inclination you might have station information at various points along the path, which will make the

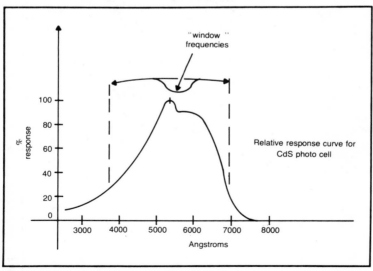

Fig. 4-3. A typical response curve for a CdS photocell.

robot do some other things when it gets to that point, such as stop, speak, raise its arms, sing or compute or whatever! Just add some "dashes" to one side of the path line and in a position where a third sensor can pick them up. When the robot base moves over that spot it gets a count of pulses according to the number of "dashes" you have drawn. If you have one dash means *stop,* two means *sing* and so on, then you can make the robot do amazing things without any apparent human control or intelligence. See Fig. 4-4 for a possible arrangement.

Now, if you want to be really fancy, think of using some optical fibers to scan the floor for the path and control signals. These would have their own light sources "built-in" close to the detector units and there would not be any large amount of radiated light to be seen under the robot's base. We suggest that if you want to make an easy control system you might use the output of the photocell or photodiode to operate a dc amplifier—there are many in the form of integrated circuits—and use the amplified current output to operate some relay or relay-type circuit to operate the motors. As shown, two drive motors are used (one on each side of the base and independently powered). If they both run at the same speed the robot's base moves straight forward or backward. If one moves faster than the other, the base will turn toward the slower wheel.

One might make a path of spaced pulses and use a pulsed motor or motors so that the synchronizing problem will not be so great.

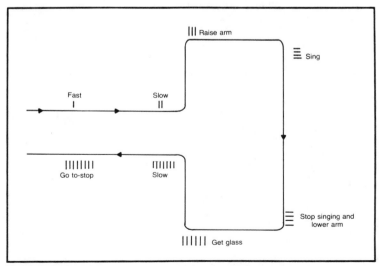

Fig. 4-4. A plan showing how to control path steering.

Each time the base passes over a pulse both motors will advance the same amount except on turns, when the side control signal can insert some resistance onto the lead of one motor so it turns more slowly and causes the steering required for turns. Yes, it will take some experimentation. The ideas are all sound and it could be a "fun project" to try to assemble and make operational. Also, there is plenty of learning and education to be derived from the effort! Good luck!

SOME SUGGESTED CIRCUITRY FOR A PATH-FOLLOWING ROBOT

At the outset let me call your attention to two books on robotics I have written, published by TAB Books Inc., Blue Ridge Summit, PA 17214. They are: *The Complete Handbook or Robotics (number 1071),* and *Advanced Robotics (number 1421).* These could be of help to you if you investigate the robotics projects. Also we recommend that you obtain some literature on all possible electronics circuits that show comparison amplifiers, digital to analog circuits, analog to digital circuits, and control circuits for steering and motor control such as those described in TAB books numbers 1174 and 1135.

Before examining some circuitry, let us also consider another "path" arrangement as shown in Fig. 4-5. In this diagram we show how you might draw short "cross-wise" lines with your invisible ink that will be fluorescent and can be "seen" by the photodetector. What happens is that as the robot passes over the line it will "see" a series of pulses. When the pulse strength and rate is equal into both

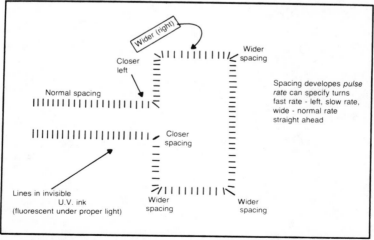

Fig. 4-5. Making a path that can generate electronic steering pulses.

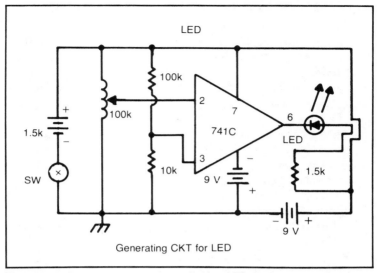

Fig. 4-6. A light-generating cirucit used with a LED.

photodetectors, then the robot base goes straight. But, if you close-space some at a turning point, then a counting circuit (which you will obtain) inside or in series with the photo output, will cause a turn-over-ride signal to be generated to cause the base to turn in a given direction so long as that fast pulse rate exists. A return then, to the normal rate (governed by how closely or far-apart you space the lines) will again release the steering control circuits to make the base go straight. To make the robot base turn in the opposite direction, you'll need to have another type of signal. This can be accomplished by using a wider spacing (slower pulse rate pick-up) and that signal could mean go in the opposite direction until the normal rate is again "seen" by the robot's eyes when properly decoded.

It can be very interesting to imagine what you can do with variations of the pulse rate, from wide (long duration) pulses, slow pulses, narrow pulses, and fast pulses to a special coding of pulses as laid out on the floor or path. We imagine that with a proper circuitry and proper coding you might be able to pre-set a multiple of commands for your robot. Once activated, it will start moving along a given path stopping at various places to do a multitude of things such as singing, or opening doors and such. The capabilities are only limited by your imagination and electronic genius. Use a path like that shown in Fig. 4-5 and anything is possible!

To get you started on such a project we've included some circuits in Figs. 4-6 through 4-9. The three circuits shown in Figs. 4-6, 4-7 and 4-8 might be used in various ways. The LED circuit (Fig. 4-6) might be used to cause a LED at the end of a small tube (isolated from a photodetector) to illuminate the floor when placed so it will be in close proximity to the lines. It will have a sharp and small field of vision.

The circuit shown in Fig. 4-7 might be used if a balanced output level is needed and this level can be affected by the light falling on the photo-detector. If the light is more intense, a larger output of some polarity might be obtained, and if the light is less than a given level, the output might change polarity and be smaller. You have to experiment with these circuits to see what they do and what you can do with them.

The circuit shown in Fig. 4-8 might be the one to use with the circuit in Fig. 4-9 to cause the robot to steer itself. Here a photo-diode will produce an output when a light flash impinges upon it. This is amplified in the two amplifier packages and sent to a transistor output that can be coupled to the input circuits of Fig. 4-9. Two such circuits are needed for steering, as you need two photodetectors. If you use a placement of "side pulses" as mentioned earlier, then you might use a third circuit like this with a photodetector positioned to "see" those pulse marks. Some component values are shown and these might be okay, but don't hesitate to change them or experiment to get what you want—and that is a voltage or current pulse at the output for each "line" segment "seen" by the detector photodiode.

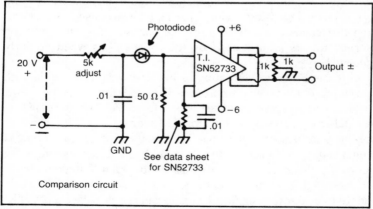

Fig. 4-7. A comparison (analog) output circuit with a light-detector input.

120

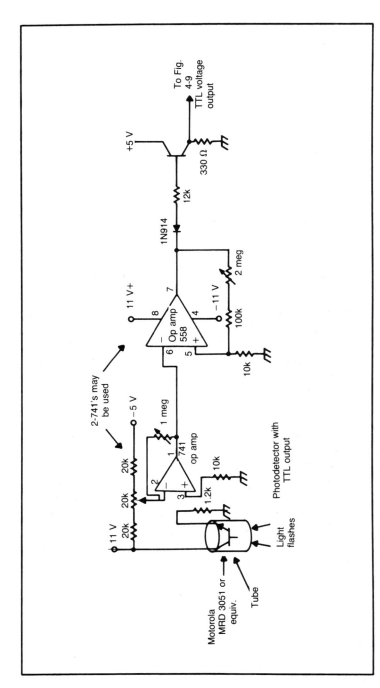

Fig. 4-8. A photodetector-to-TTL type output.

121

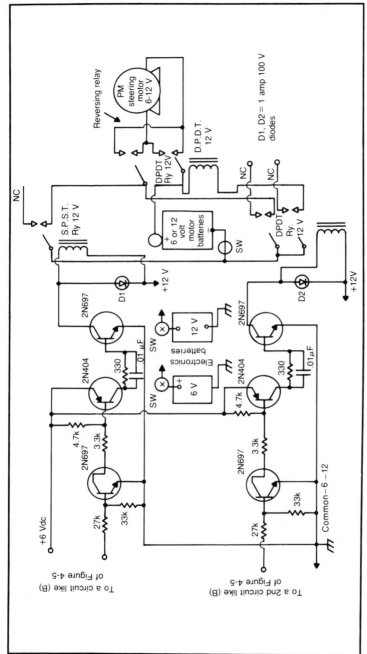

Fig. 4-9. A possible steering circuit for a path-following robot.

Finally, some kind of steering circuit is needed. We show a possible one using relays of the 12 volt type which can be transistor activated. We also show some types of transistors but others of an equivalent type certainly can be used also. The idea is simply to take the output from Fig. 4-8 and feed it into this circuit in such a manner that an appropriate relay will close and hold closed as long as the input continues at a given level. We do not show any "holding" or delay-type relay arrangement. You might need to modify the circuit to get this to hold over the pulse spacing. A large capacitor across the relay, or some other type of delay-in-releasing relay circuit might be used.

The operation of the circuits would be like this: The robot is placed on the path so both detectors can "see" the reflected light pulses, or steady stream of light rays. The circuitry will compare the intensity of the reflected rays and if they are the same (*and below a maximum level*) the robot will go straight. If, the intensity to one detector gets larger and so its output gets higher, then that part of the circuitry that causes the steering motor to activate and turn the robot's base so that balance is again accomplished, will be activated. You have to be sure the robot's base turns in the right direction, otherwise it will move entirely away from the path!

These are some suggestions and hopefully will form a foundation for your experiments and accomplishments. It won't necessarily be easy, but whoever said anything worthwhile was easy? Your satisfaction seeing your robot "run-around" doing all those things you specified for it to do, at the places you specified for it to do them, will be an utter amazement to your friends, family, and anyone else who witnesses the performance. So, if you have the time and want the challenge—try it!

MORE HINTS FOR THE PATH-FOLLOWING ROBOT

As ideas occur it is helpful to write them down. It is also helpful to "doodle" with a pencil when ideas occur to you that result in possible circuitry or construction of a path-following robot. One such idea just blossomed in our mind. It would be nice to have a path for testing purposes which wouldn't require that we paint white strips on the floor, or put invisible ink strips on a rug or such. So we offer this idea for what it is worth to you.

Why not purchase a long roll of wrapping paper that is a dark brown and wide enough to form a path when properly marked. Being paper, it will be easy to paint white strips on it or put any other color marking which you might want to try for a robot sensor to detect and follow. Then, when you want to have your robot machinery operate and you test it to see if it will actually come anywhere near doing what you want it to do, you simply stretch out the path segment (any length you choose) and place the robot on the paper with its marked path and turn it on to see what happens. Your family, especially the lady-of-the-house will love this idea because it doesn't mess up her home arrangements. You simply roll-up the paper when you've finished that experiment and you can use it again and again as needed later on. Also, it works well inside or outside the house, on a porch, in a garage, on a sidewalk or wherever! Figure 4-10 illustrates this idea.

SENSING THE LIGHT REFLECTIONS

In your experimentations you may find that fluorescent lighting may not work well. Instead you may want to develop a special light-emitting probe with a closely built-in sensor of the phototransistor type. One such type of phototransistor is the FPT 100 or Fairchild FTK0031. It was said in some literature that these phototransistors work best with a standard incandescent light or with an infrared emitting LED. Probably these phototransistors will work all right with something like an automobile light that operates from a battery—it is worth trying. In any event, one might make a probe such as illustrated in Fig. 4-11 and see if it won't pick up the light flashes or reflections from the path marking well enough to produce

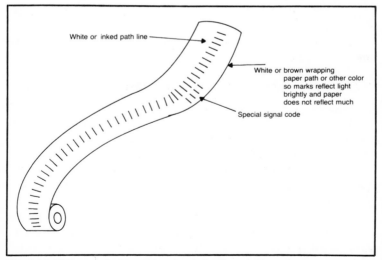

Fig. 4-10. Using a roll of wrapping paper for the robotic path.

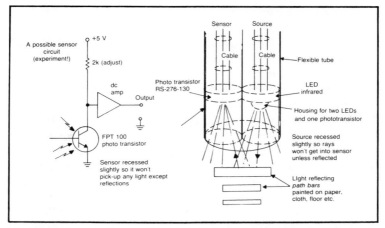

Fig. 4-11. Concept of a possible phototransistor-LED path probe for a path-following robot.

some signals to be used for steering. As you might imagine, a path using this type source and sensor arrangement could be very small indeed!

TRYING TO GET THE ROBOT TO OPERATE

Once you've got the equipment so that pulling the paper path under the sensors will cause the steering mechanism to move the wheels fast and slow or turn them, then it is time for a trial run. We suggest that you adjust the path for a "straight run" first. Just see if the robot will follow a straight line and be able to steer itself to stay on that straight line or path. *Then,* if that turns out okay try a gentle turn and add some extra control signals. Developing the system a step at a time until it all works just fine and you are ready to give the world a demonstration of your brainchild. Oh, what fun! By the way, the nice thing about using a sensor that operates on infrared wavelengths is that the sensor can be adjusted so it is not sensitive to ambient light, so you might be able to demonstrate the system inside or outside in broad daylight!

Then, if you are really experimentally inclined, you might use fiberoptics to bring the light down to the path and the reflections back up to the sensor that can be placed at the best possible location in the electronics package. Also, of interest is the fact that in many applications of optical fibers multi-strands all packaged together are used to form a "bundle." This makes it more redundant in reliability, as if one strand should fracture there will be many others to transmit the light rays over the cable. This is another important idea to consider.

When we are handling optical fibers which are so terribly small in diameter, it is nice to have enough of them grouped together so that we have something "physical" to grasp hold of. So a nice cabled-packaging of fibers might be made up to a diameter of some ⅛ inch or perhaps under some circumstances to ¼ inch in diameter. These are all independent but they also work together to transmit light rays from one point through themselves to another point.

Not the least important of the various items needed for such light transmission would be the connectors. These connect the ends of the optical fibers so that the light rays can be brought together and passed into the device that receives the rays. We suggest contacting Oriel Corporation, 15 Market Street, Stamford Connecticut 06902 for their catalog and information on any optical needs you might have. In Fig. 4-12 we show a sketch of how an xenon or a mercury vapor arc lamp might be coupled into an optical fiber transmission line using some of the special lenses available from Oriel.

There are many types of light sources and these will produce various frequencies of visible light. So, you determine the type of light desired (infrared, ultraviolet, or whatever) and then you try to get a light source rich in that frequency content. You also get an optical fiber that will pass in a mono-mode (hopefully) the frequency being generated. All this takes some consideration, investigation and planning, but the results are worthwhile.

SOME VARIOUS OPTO-ELECTRONIC CIRCUITRY

Since we have turned to somewhat of a practical approach in this chapter, let us continue the investigation by looking at some

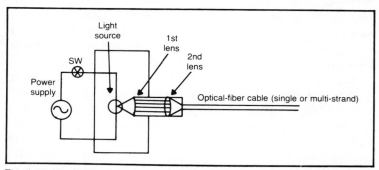

Fig. 4-12. Illuminating an optical fiber with lenses and a light source.

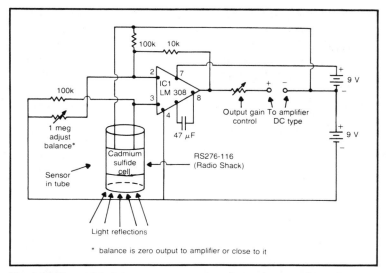

Fig. 4-13. A possible light-sensing circuit using a cadmium-sulfide cell.

various electronic circuits that use optical-coupling or light-coupled circuits (both being the same thing). One of the first types of circuits (and an effective one) uses a photoresistor element. One type is the cadmium sulfide photocell (Radio Shack 276-116 or equivalent). This cell will change resistivity when light falls upon it. It can be used in almost any circuit using a resistor arrangement as input and gives an output proportional to the light intensity falling upon it. It works with a wide range of light types, from flashlight to ac bulbs. When used in the circuit shown in Fig. 4-13 it can control other devices.

The circuit is a variation of a Wheatstone bridge. This resistance cell is used as one leg of the bridge. The one megohm balance resistor does just that—it balances the output to zero by balancing the bridge input section. The gain control will govern the current flowing through the output terminals as the input becomes unbalanced. Thus a transistor connected properly across the output terminals can easily amplify the current input and give a good healthy output current (use a Darlington if necessary) that can cause a steering motor to run or turn or whatever.

Now, if it happens that you can generate enough current to cause a LED to light, you have enough current to operate an opto-isolator circuit. This comes in a small integrated circuit package and it consists of a light source and a receiver in this small

127

package. If the output from your circuit causes the light source to produce enough light then you will get an output from the receiver section which could be a voltage, current, or TRIAC control of a large ac current. You will no doubt investigate all these types of opto-isolators at Radio Shack or a similar store to find one that will operate satisfactorily with your other circuitry. The advantage is not only circuit isolation (separating the sensing circuits from the motor control circuits to do whatever you need to do with a motor which is so-controlled) but also control of a much higher current from a secondary source battery supply.

Another circuit which might be of value is that shown in Fig. 4-14. The ingenious experimenter will be able, possibly, to replace the motor with a small resistor, and use the voltage developed across it to feed some kind of transistor amplifier (that could be direct-coupled) in order to get a much larger current control for, say, a much larger motor of this PM type (permanent magnet). But, that investigation we leave up to you. One suggestion, however: if nothing else, a small relay with a low resistance coil might be substituted for the motor shown, and that relay through its contacts might be able to control a much larger motor's current. Have a "go" at it!

THE SCIENTIFIC METHOD IN EXPERIMENTATION

Among the "gentry" there is always much hinted and stated with regard to the "scientific method." This is a step-by-step procedure recommended for experimenters and those investigators whose skill level and knowledge level places them a little beyond the ordinary experimenter. In other words they seem to have some firm idea of what they are after and they devise experiments, somewhat skillfully (sometimes) to prove or disprove certain ideas. So we state the *scientific method* for what it may be worth to you. It is: Conceive the idea or theory. Build the equipment to conduct the experiment to prove or disprove the theory (this may take some doing and some delving into design concepts, and even development of new types of devices and machines and such to be able to conduct the experiments). Relate the theory to the experimental results by conducting the experiments. Then publish the results in some prestigious scientific journal.

What is *not* stated in the scientific method process, is that one may have to conduct a million and one experiments in order to verify or *prove conclusively,* positively or negatively, that the theory or idea is true or false. If true, then one has to come up with equations

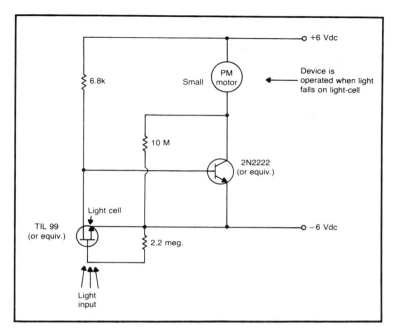

Fig. 4-14. A simple light-activated control circuit.

that permit others to duplicate the results obtained. Also, one probably has to be prepared to *defend* the innumerable attacks that always come upon one's experimental procedures. Are they complete enough? Are they valid enough and so on? It is not stated, but somehow taken for granted that you know that an experiment has been proved when the results from countless experiments, under different conditions, may produce the same results. If there are variations in the findings then one goes "berserk" trying to figure out why!

Including this bit of information in this chapter is done to alert you to the fact that with lasers and optical fibers there is much to be learned and perhaps fabulous results might be obtained from such experiments as you might undertake. Recall that some of the most "explosive" developments have been said to have been the result of "accidents." It was said that the development of gun-powder was an accident resulting from an investigation of chemical combinations for some other reason—perhaps alchemy. Who knows what you might come up with. Be prepared to write down small events and happenings and things which are either good or bad or useful or apparently not useful. Recall that at least one attribute of the good experimenter is that he keeps a notebook handy and notes every trial and effect almost minute-by-minute. He doesn't wait till the end of the day and then try to recall everything to write down. Finally, he puts his investigations into a good form on paper and submits the manuscript to a suitable publisher.

OBTAINING SOME OPTICAL FIBERS FOR EXPERIMENTATION

So, the next question is: "How can I get some optical fibers to do some experimentation with?" The answer, we suggest is to write to Edmund Scientific Company 7785 Edscorp Building, Barrington, New Jersey 08007 and ask for their catalog in which optical fibers are listed, also information on optical fiber experiments that you might perform (also fiberoptics kit # 70,855). That, at least might be a start. Then, for lenses and light sources and such, contact the Oriel Corporation 15 Market Street, Stamford, Conn. 06902. This effort may give you the materials you need to begin your investigations.

Edmund Scientific has some light fibers bundled into cables so that from 16 to 64 such fibers are sheathed in a plastic covering and can be used in experimentation. What experiments? Well, vision—that is, looking into one end and seeing what is near the other end for one thing. You can bend these fiber cables around and still easily see what the other end "sees" *within its acceptance cone of some 70 degrees!* It has been suggested that you might look into walls, into machinery, into people, or whatever and see what is going on in there. You cannot do this normally because you have no way to get the reflected light from the inside of things to your eyes, which are on the outside. So, you might use one cable for a light source, sending light rays through it into the darkness you want to investigate and use a second cable to actually look through to see what is there.

What a nice idea! It occurs to me that in some homes, behind the sheetrock walls (see our book *Building, Remodeling and Renovating Homes, TAB book number 1287*) that you might want to look at electrical wiring, or plumbing, or see if a rat's nest is located there. You don't want to tear out the wall, or cut a big, unsightly hole in the wall to look around, so you use optical fiber cables. These take only a very small hole to get the "light source" cable and the "seeing" cable inside to look around. These holes are easily patched if nothing is there. If something is there that needs correction in some way, *then* you can open the wall up for access knowing that you aren't just tearing out some wall for the fun of it! *Edmund Scientific* calls their light fibers *Crofon* which, they say, transmits light uniformly in the visible region, reduces transmission sharply in the ultraviolet region and absorbs most of the light in the infrared region. The material is relatively strong and reasonably larger in size (a single fiber being some 10 mills in diameter). To give you some idea as to the relative size of the cables they produce see Table 4-1.

Fittings for an eyepiece will be required for the cable used to see through, and light sources will be needed for the illuminating cable. It is to be supposed that if one is adept at experimenting with these fibers, that one might take a multifiber cable, split the fibers at one end, use some for light input and some to see through. That would also make a fine kind of sensor using very fine lines wouldn't it!

USING THE OPTICAL FIBERS IN EXPERIMENTS

So, what do you do when you get an optical fiber cable? You first cut it to your required length using a sharp knife, a pressure cutter such as you have seen that trims paper sheets (a sharp razor blade has been suggested). In any event you will want to get the cable cut to the right length for your use. Second, you need *to polish the ends* so that the best light transmission will be obtained. It is said that you can do this with a special lacquer (we cannot venture more information than the fact that this material probably exists) or by some abrasive polishing method. You just get some kind of abrasive (such as wet#600 alundum), and being careful to hold the end of the cable, which you have cut perpendicularly to the cable length, against a very fine grinding wheel, polish with the required motion. You might polish by using a wet sanding with 400 grit and 600 abrasive paper, then finish the polishing by cleaning thoroughly and polishing with a cloth wheel or chamois.

By the way, if you get some liquid lacquer, the way you use this is to cut the end of the cable carefully and perpendicularly, then dip the end of the cable into the solution and let it dry for some 10 to 15

Table 4-1. Edmund Scientific Company Optical Fiber Cables.

Stock number	Fibers	Outside diameter (inches)	Cross section (square inches)
2504	16	0.087	0.001
2505	32	0.110	0.002
2506	48	0.119	0.003
2507	64	0.130	0.004

minutes. Edmund Scientific has the lacquer and sanding materials for their Crofon optical-fiber cables.

ADDING THE END FITTINGS

These are put in the cable by a close-pressure connection that may use heat shrinking plastic wrappers to hold them in place or by metal pressure connectors or by gluing with special epoxy that is clear to the light rays after drying, and so on. You might want to add lenses to the ends of the cables and such lenses are being made and are available. You might want to add filters of various types to limit the type of light (frequency or wavelength) that is propagated through your fibers. Light sources that have been used with Crofon optical-fibers range from Tensor lights to an ordinary light bulb, to lasers and LEDs. **We caution again about the light: Be Careful, especially with laser light. Don't take chances with your eyes. Use smoked goggles. Don't look directly into the light source, or allow anyone else to do so! Use any other precautions you can find out about to prevent damage to your eyes or those of your family and friends and colleagues!**

Finally, we show a sketch of the acceptance angle of light for a *Crofon* cable optical-fiber. This is interesting in that the acceptance angle is much larger than we have heretofore been led to believe it might be. Examine Fig. 4-12. It probably needs to be stated that only is the light-acceptance angle as shown, but the illuminating angle for light emerging from the fiber and the visual "seeing" angle are the same as that shown. It's a good wide view and makes light input relatively easy.

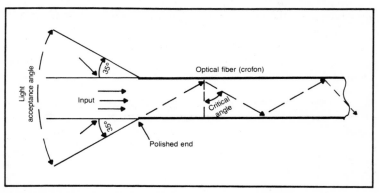

Fig. 4-15. The light acceptance angle for Crofon optical-fiber.

LIGHT DEVICES FOR COMPUTERS ARE NOT LIMITED TO FIBEROPTICS

The search and research goes on looking for better and better "mousetraps." We find a report from the Hughes Aircraft Corp. that says their research department has developed a new all-optical *logic* device which is bipolar. It is said to consist of four reflecting surfaces and a slab of nonlinear gallium-arsenide and it provides high speed switching in a unit that is impervious to any random or constant electrical interference. It is said that the device will "flip-flop" reliably with a switching time of as little as 3 nanoseconds and switch energies under 100 microjoules. They want this unit to come near the theoretical switching limit of 10 gigahertz/second so it can be used more effectively in computers and flight control systems and signal processors. A research effort is being made to still further reduce its size so it can be contained on an *optical integrated circuit chip*.

What could a bipolar switching device be used for? It operates on a light ray or rays. If you can imagine a window shade which can be opened and closed a billion times a second you have an idea as to what this switching device *may* represent. We know that previous research has produced some devices which can block the transmission of light rays when a given size and amount of electric potential is applied to them (a crystal of special type, for example) and so when such a current is removed, this device then permits passage of the light rays. This is to all effects and purposes a bipolar switch. This switch can be held in either state (on or off) for as long as the applied potential exists or does not exist. It is a controlled and reliable bipolar device. The Hughes device may not be just that but it may act in the same manner. As we have said; the search goes on for "better mousetraps."

USING OPTICAL FIBER CABLES IN AUTOMOBILES

Well, of course you are familiar with this concept. You simply terminate a cable near a taillight and run the cable (along with other wiring in a larger cable) to above the rear window. Then you put a cover on the tiny compartment with two holes in it so the ends of the optical cable can go into each hole. When viewed from the front seat (via the rear-view mirror) you can clearly see the little bright dots which show you that the taillights are on, or that the taillights are off. Activating your turn signal will make one little light blink so you know that the system is working.

You are probably familiar with the front fender light dots which show you when you have parking lights on, regular headlight beams on and/or the distant illuminating beams on. This is a handy little extension above the front fenders over the wheels, not high enough to be really noticeable, but there to show you that all your headlight beams are working properly. Fiber-optic cables connect these to the proper light element in the front of the car. You do not see any cables showing, nor do you have to worry about electrical problems or deterioration because the fiberoptic cables need no more energy than the light which goes into them. But they do produce results and are handy to have.

At least one company, *Hitachi,* has done some study on just how effective would be a replacement of the electrical cabling system in a car with an optical fiber system, capable of translating the current-voltage demands from one area to another. They suggested that laser light emitters might be used to transmit untold numbers of digital signals through optical fibers to suitable diode-type receivers to monitor functions and operations within the car itself and present this information for display or use the information for better automatic adjustment of the various devices that make the car run.

Their study indicated that their researchers used some LEDs that produced light in the infrared region and sent it through very small diameter optical fibers (1 mm diameter) to suitable light receiving diodes. They are of the firm belief that the entire automotive electrical system might be replaced with such a system. The gain would be better efficiency, lighter weight, probably less cost in the long run, and less interference generation and susceptibility. The fact that computerization of automobile adjustments to the engine are increasing all the time and are being made more and more automatic and autonomous, means that reduction of electrical interference is a primary requirement if the complex computer devices are to operate reliably and satisfactorily. Perhaps, in the future, such devices will permit automatic steering and control of your automobile, once it is on a properly instrumented highway. How nice! You just jump into your car, set the computerized control for "office" and sleep, read the paper, listen to the morning news, or eat breakfast (of course breakfast is prepared on the spot inside the car in a special compartment). And finally, the car stops in its prescribed parking slot at the office.

MILITARY CONSIDERATIONS OF
OPTICAL-FIBER COMMUNICATIONS AND CONTROL

"Communications silence!" This expression indicates how important it is to be "secure" in the emission of electromagnetic radiations from ships, aircraft, tanks, vehicles, or other components of the military operational establishment. To transmit any information by normal radiation means could result in telling the enemy just where you are and could be providing the point-source information to accurately steer an electromagnetically-guided homing-missile or rocket of some type directly to you. On shipboard we think of optical fibers as a base communications system or connecting system between all the ship's elements for communications, control, computation, assessment and all other required functions. This connecting system, while having all the advantages of providing the necessary functions and information does not radiate!

A LIGHT-OPERATED COMPUTER

In this concept it is informative to know that at Ohio State University in 1981 an all-optical digital computer was demonstrated. This computer used a Hughes liquid-crystal light valve. It is said that this device accepts optical images (light rays) and replicates them on a completely separate light beam from an arc lamp or laser. The technology is similar to that used in the fabrication and operation of liquid-crystal watches. But the important thing is that *optical* equivalents of electronic logic gates and flip-flops have been constructed with these "light valves." They perform in much the same manner as do the transistor logic and gate circuits of the electronic computers. In an optical computer the computing operations are said to be much faster, the computer more immune to interference and, thus perhaps, more reliable. This "light" computer uses photons (bursts of light energy) in the same manner that the electronic computer uses electrons. Perhaps connections to the various sections of the computer is accomplished by optical-fiber connectors and cables. That also seems logical. Thus we may yet see a revolution in computer development in the years ahead. Optical systems have the main advantages of security (non radiation), and wide bandwidths, which could mean a much higher rate of data transference than would ever be possible with wires and electronics as we now know it.

MODULATING A LIGHT BEAM

At the Naval Post Graduate School at Monterey California, an optical communications system was constructed which a pulsed gallium-arsenide injection laser was used for the transmitter. It operated at a repetition frequency of 8 to 15 kilohertz using a pulse-position modulation concept to convey intelligence. A PIN diode was used in the receiver. The pulse position modulation was used to transfer information as well as to trigger the laser so the light beams produced had the same pulse pattern.

REMOTE CONTROL USING A LIGHT BEAM

Of course it has to be true that someone would investigate the use of light beams as a means of conveying commands to a remote receiver for remote control of various devices and operations. The use of a laser as the transmitting source for this kind of experiment has been accomplished. The receiver (in order to be the equivalent of a broad-band conventional super-het) has to be a type which looks everywhere for the incoming light flashes that represent commands. The receiver has a 360 degree field of vision. Such a remote control system, without isotropically radiating energy become secure and in most directions silent. It can be used where no other type of control system might be employed for whatever reasons prevent their use. In the reports it was said that a continuous wave helium-neon laser served as a transmitter and that its output was suitably modulated to convey the command instructions.

In the field of information storage, it is perhaps amazing to learn that optical-storage systems offer an almost unbelievable capacity to retain digital data. A single optical disc, which is recorded with a special laser beam, and is the size of a long playing record, is said to be capable of storing *billions* of digital data bits. Just how much data it stores depends upon the size and spacing of the laser produced holes in the disc, and also in the type of coding used. It has been demonstrated that from 500 million to 50 billion bits can be stored on a single side of such an optical-storage disc.

Some of the advantages of optical-storage turn out to be the longer life of the disc as compared to the electromagnetic type of disc-storage. But there are also disadvantages and one of these is said to be the inability to erase and change data on a given disc. It is said that the error rate should not be less than 1 bit in 100,000 bits, and that is a pretty high factor. Those who use computers say that this rate is not good enough, they want only one error in 10 trillion bits transmitted or stored!

Of course we know that optically written data can be written on a credit card or other flat type "instrument" which is used by the general public. The information stored is not only secure, but also may contain much more information than anyone can imagine should be contained in such a small area or space. The mechanics by which data is stored is said to be in small pits or holes which are laser produced, and this then, produces a different reflectivity to an impinging light ray, than the surrounding material. Thus, if there is a good reflection, this might mean a "one," while a pit, causing a much less bright reflection might mean a "zero." A "pit" or hole may be only one micrometer in diameter and that takes a small size laser beam with some good optics to get the scanning job done. One supposes that the laser is modulated, turned off and on by a pulsed input signal so it creates the pits and blank spaces between them in precisely the same relationship as the modulating signal information. Thus the "information" is stored.

Now, to "read" the data requires some doing. First there must be an appropriate light source, another laser which has a pin-point beam, then some *fiber-optical cables* to convey the laser output to some beam splitting lenses, and these will split the beam so that one half goes to the disc to be reflected back. The second half of the beam is used to generate a reference signal so the electronics will know whether the reflected beam is of full intensity (meaning no hole or pit) or is partially absorbed in the hole or pit so it has a lesser intensity. These two extremes of reflection can then be interpreted by the electronics detector and comparison circuits as being ones or zeros.

Of course, there must be a tracking system which causes the laser beam to follow the "tracks" of data on the recording. That is another complex electronics system, but one which has been developed and works satisfactorily. Also, one then can imagine the complexity of the system to "go in" to any portion of that disc to extract a particular piece of data! Drexler Technology makes a material for optical recordings and they call it Drexon. RCA is also investigating this field of optical storage as well as other known and respected companies such as 3M, and Thompson-CSF in France, and IBM. This area of optical-storage development is growing and it will cause some changes in our concepts of computers and what they can and cannot do. Just imagine writing 450 million bits per second! One day maybe we'll find out that our own human brains are too limited and too slow in operation—what do you think?

USING OPTICAL FIBERS IN DIRECTIONAL REFERENCES

It has not been too long ago that the old mechanical gyroscope was the standard reference system in guided missiles, aircraft, ships, and submarines. It is still used because it has proven reliable and rugged and accurate. But new developments (better mouse-traps) are always forthcoming and so modern technology now acclaims the "laser gyro", which is a rotational rate sensor system which can be used to roll-stabilize missiles, and aircraft and (using computers with good memories) determine positions and directions and such.

The "laser gyroscope" operates on the principle that you can split a light beam (coherent type) so that if you send it along two different paths, and these paths are in the perpendicular plane to the direction of rotation of some body, then the light beam in the path moving in the direction of rotation will move faster than that which moves along a path opposing the direction of rotation. How do we know this? We measure the phase or frequency of the two beams after they have traversed the paths, and we find the phase or frequency different. Of course, it is assumed that we start with a single beam, which is "in phase" or coherent and split it to make up the two beams used (so that one may act as a reference to the other).

Well, anyway, the beams are light and so much effort and research went into the "proper paths" that the light beams from the laser should follow, and then the researchers found that it took some distance and that means big devices. Not exactly what they wanted. They looked to optical fibers, whose elements could be wound up on spools to make nice long paths and since the fibers are small diameter (and wind up into very small spools) they got the small sizes they needed. Aha! Long paths and small devices—what more do you need? Again, the optical fibers assume a most important role in a very scientific device.

Everyone was pleased at the results demonstrated by Victor Vali and Dick W. Shorthill of the University of Utah who first demonstrated the practical construction and operation of such a device using optical fibers. These gyro units have a very slow *drift* rate (meaning the shifting of the reference) of about one degree per hour and this compares to some mechanical gyroscope's drift rate of up to 5 degrees per hour in some cases. So, with the small size and accuracy available what is the "trade-off" (the penalty of use)? Some say it is the difficulty of getting the laser light beam into the optical fibers since the laser beam is a relatively large diameter and the fibers are a relatively small diameter. Also, it is said that with the optical fibers there is still some accuracy problems. The optical-fiber gyroscopes still are not as good as just the ordinary, larger laser gyroscopes using the same principle of operation. Some say that the optical-fiber gyros are in the category of mechanical gyros (good ones that is) but improvements are on the way.

There is a need for much experimentation and development in optical-fiber connectors, light sources, transmitters and receivers, polarizers, and phase splitters and such. Also much development is underway and will continue to produce optical fibers with little loss, wide "windows" and monomode operation to reduce dispersion losses. Just as we went from hard-wiring in electronics circuits to the printed circuits and then to the integrated circuits—so we now progress to optical-fiber connected elements and the use of light as the active element. It is a normal and expected progression. Get involved!

USING PREFOCUSED LEDS

Some firms are making prefocused LEDs for use with optical fibers. One such firm is the Honeywell Opto-electronics Division of Dallas, Texas. The lens becomes an integral part of the glass or plastic envelope of a LED as shown in Fig. 4-16. It is said that these plastic LEDs are low in cost, and can be used with optical-fiber cables that could link together the home computers, TV sets and cable television distribution systems, and the computer ancillary equipment such as printers and files and disc drivers and such.

In the manufacture of the LEDs using a plastic covering, a lens is made right into the end of the unit and it may be spherical in shape. This is a somewhat different shape than we have thought of as a lens (or focusing element's shape). However, it is not unusual.

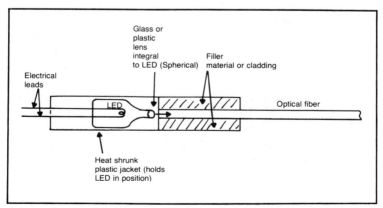

Fig. 4-16. Using a prefocused LED with an optical fiber.

As mentioned previously, light rays are an extension of the electromagnetic spectrum and the rays behave much like radio, radar, and other ultrahigh frequency radiations. The military has used focusing in their spherical Luneberg Lens antenna systems and there is no reason why the small glass bead of spherical shape would not focus light rays in the same manner in which a Luneberg lens will focus radar waves. By the way, the concentration point of light in a LED unit, when focused, is called the "sweet spot" so we are informed.

MORE ON THE ROTARY MOTION SENSOR USING OPTICAL FIBERS

This device is a rotary motion sensor which employs an endless closed-loop multiple-turn optical fiber interferometer. Coherent laser radiations are coupled into the (optical) fiber interferometer in both a clockwise and a counterclockwise direction and the energy is permitted to traverse the entire light-waveguide many times to increase the accumulative phase shift which will occur due to the multiple transversals. There has been devised a means to get the laser signals into and out of the light waveguide. Also a means has been devised to generate a pair of reference signals which are then hetrodyned with the clockwise and counterclockwise sensor signals to produce a pair of frequency signals at a "manageable" frequency. The relative phase shift between the acoustic output frequencies and the circulating signals is the same, so they provide an extremely precise measurement of the rate of rotation of the device. In a preferred configuration, the means of coupling and the means of reference signal generating comprise an acoustic waveguide for generating a required opto-acoustic grating across the paths of the coherent optical signal and the optical-fiber interferometer. The detection of phase differences between reference light signals and sensor light signals would seem to require some kind of diffraction process.

It has been found that the sensitivity of such an optical-fiber system to inertial rotations is adequate when the loop consists of some 600 turns of single-mode fibers. In the system tested, the name Sagnac was used as identification. Also much work was done on light amplifiers which could compensate for the losses in the optical fiber path. When such a system can accurately detect the very slow rotations of objects with respect to inertial space (which is assumed to be fixed in rigid coordinates) then the uses of such a rotational sensor will be increased many fold.

WHAT ABOUT AN OPTICAL DELAY LINE?

The delay line is often used in Early Warning Radar systems to preserve the frequency and phase relationships of rf signals of the microwave frequency range. Such delay lines have to meet very rigid requirements such as low insertion losses, wide range and bandwidths, and produce as nearly as possible a linear response. If we convert the microwave energy to light energy and then use an optical-fiber delay line we find that, again perhaps a spool of the optical-fiber material is used. It takes time for the energy to traverse such a spool and thus "delays" are obtained. What is needed is a single-mode type of fiber and a high speed photodetector and rf amplifier. Studies have been made of such delay lines operating in the 4.0 to 6.5 gigahertz range, and direct modulation of the light signal is accomplished with the use of an injection laser.

OPTICAL-FIBER COUPLING DEVICES

A patent has been granted which discloses a portable unit that can couple optical fibers together so energy will transfer among them. This unit consists of an insulative block to which the optical fibers are attached. A piezoelectric block is attached to the fibers also and connected to a source of energy. When electricity is applied to the piezoelectric element, an arc is developed across the electrodes in such a way that the optical fibers are heated and caused to flow together, joining them. The coupling is solid and tight and as good as a couple can possibly be made. It is better than just pressure-joining the polished and flat ends of pairs of optical fibers or optical fiber cables.

AN OPTICAL-FIBER MECHANICAL SENSOR

The US Navy Department has invented a mechanical sensor that uses an optical-fiber in a sensory role. it is said that this concept uses an optical fiber to modulate the intensity of light passing through an optical fiber in response to mechanical motion in the form of linear position, rotational rate, or strain on the fiber. The light from a suitable source is sent down an optical-fiber path through a birefringent modulatable element which is positioned between cross-polarizers and then that light goes on to a suitable diode detector. The birefringence of the modulating element is susceptible to the influence of an external source of pressure such as a mechanical stress movement or strain, and also it is susceptible to change due to an applied magnetic field that can be generated by the device being monitored or "sensed." This type of external force causes a rotation of the light in the modulating element and thus encodes the information when there is a mechanical change, in a kind of digital signal that then can be detected and decoded by suitable electronics and computers.

We know that when light passes through an optical fiber it can be weakened or dispersed by small anomalies in the fiber such as nonsmooth surface effects of a winding spool. Thus it is not beyond our understanding to realize that if we can subject the optical fiber to any physical stress or strain that some change in its light-conductivity will take place and that we can detect this change in light transmission and relate it by some mathematical algorithm to the physical changes in the sensed element. Then we have a sensor and a good one at that!

Earlier in this text we described birefringence. You may recall what it is. In case you do not, we again offer a small description. If we take a crystal, such as calcite ($CaCO_3$) and lay it over some printed words on a page, the image we see will appear double. That is called double refraction. This is due to the formation of two beams of reflected light with different polarizations with respect to one another. One ray of light will be refracted in the normal manner and obey Snell's law of refraction. The other beam of reflected light does not obey this law, but passes through the crystal as though it were not there. Thus, two distinct beams are formed by this phenomena. By the same token, or phenomena, when light is passed into such a crystal, two beams emerge; one is a direct ray and one is a refracted ray. They will occupy two distinct positions in space. The fact that two images or beams are produced is called the phenomena of *birefringence*.

We have mentioned the use of an optical fiber as a sensor, so let us continue with this idea. In connection with previous developments, the Navy Department has produced a mechanical shutter effect for use with light rays and optical fibers. A mechanically actuated piezoelectrically operated, fiber-optic connected light valve (shutter) can be utilized as a mechanical motion (pressure) sensor or displacement sensor. The piezoelectric element responds to pressure to generate an electrical signal (we know that this is done in transmitters of all types that use crystals of the piezoelectric type for frequency control) that is applied to a *Kerr-cell* containing nitrobenzene. This electrical signal then makes the nitrobenzene transparent (it was not transparent to light previously) and so the light rays can easily pass through this Kerr-cell and through the conducting optical fibers to a light detection receiver where the light intensity is then converted into a proportional electrical signal of the digital form, or a coded signal of the digital form.

Of course, optical fibers are used in hydrophones as sensors. An acoustic wave detector is placed in such a device and this detector is formed from an optical-fiber coil. Since acoustic waves are pressure waves, then changes in pressure on the coil can be detected and so a hydrophone detector unit is formed.

DEVELOPMENTS NECESSARY FOR OPTICAL FIBER VIABILITY

Perhaps one of the most important areas of development to increase the use of the optical fibers is that of developing the source light in small integrated packages and developing light detection-conversion receivers in very small packages so that they can be integrated into the ends of optical fibers. Already, light sources are being constructed using small lenses and lasers and LEDs are being complimented by receiver units which can produce linear outputs over wide bandwidths. Of course it is desirable to have the optical fibers operate single mode, if possible, to reduce dispersion losses. Integrated optoelectronic devices, using silicon substrates, and single and multi-mode optical-fiber coupling systems are needed. New injection lasers, studies to provide more understanding of infrared physics, photon excitation, optical-fiber linkages, and computer-controlled laser systems as sources, are all under investigation at present.

OPTICAL-FIBER CONNECTORS

As you have already probably considered, a problem could develop using optical-fibers if they are rigidly connected to terminal devices. This presents somewhat of a dilemma. If you don't have the optical-fiber tightly held in a terminal connector, it might slip and work loose and reduce the transmission of light rays. If it is very tight and secure, then any twisting or bending of the fiber might cause a fracture and render the "light-guide" (as it is sometimes called) inoperative.

A government invention seems to offer one solution. This invention relates to an apparatus for attaching an optical-fiber cable to a terminal using a snap-ring to achieve a rotatable connection wherein one jaw of the device supports the terminal and the second jaw supports the ferrule and an intermediate third jaw supports and compresses the snap-ring. With the snap-ring initially mounted on the terminal, closing the jaws together compresses the snap-ring so that the terminal can be fully inserted into the ferrule until the snap-ring is captured within the ferrule and thus completes the connection.

OPTICAL-FIBER MEASUREMENT DEVICES

As with radar microwave guides, a system of measurements is necessary in order to determine the efficiency of the light guide when using optical fibers. An invention which described how a light guide can be measured using a combination of optical properties was developed by the Department of Energy (no longer a Department) in Washington DC. A polarized light pulse is injected into one end of the optical fiber. Then the reflections from discontinuities, which are *not* polarized are adjusted to impinge upon a light detector, whereas the polarized incident light pulses cannot reach such a detector. Thus, using the intensity of the detected reflections, the transmissibility of the light fiber can be measured.

USE OF OPTICAL FIBERS WITH PUNCHED CARDS

Of course this is now almost "old hat." Groups of optical-fibers are so arranged that they cover the total area over which such a card might have holes punched in its surface. The card is presented to the fibers with a back-light source. Where there is a hole, light passes to the optical fiber. Where there is no hole, light is blocked. The resulting light-dark combinations can then be converted into an alphanumeric display, print-out, or whatever. The interesting concept here is that a very complex and close-reading grid can be established using optical-fiber ends and then the data (light pulses) can be converted into electrical binary pulses some distance away in some electronic device and used for whatever purpose is desired (even a check of hole-punching accuracy).

CONSIDERATION OF THE PHOTODIODE RECEIVING ELEMENTS

InGaAsP APDs have been fabricated with uniform high-speed gains, perhaps higher than 42, and with quantum efficiencies of over 70% at 1.28 micrometers. It is noted that at high gains there is increased leakage currents which can produce excessive shot-noise and gain-saturation. Much effort has been spent to try to reduce this bad effect. One method is to develop *guard-ring APDs* but the results have not been conclusive. Much more effort is underway to improve the light detector capabilities, speed of response, bandwidths, linearity, and such, which are essential to the further application of light rays in computing systems. Noise generation as indicated here, is a real problem, with high gains.

SOME FANCIFUL USES OF OPTICAL FIBERS

Since the optical fibers can be used as sensors and can transmit light which falls on their ends, it seems that it should be possible to devise a chess set which could use optical fibers going to the various squares. This would convert the light-dark sensed changes to electrical signals that could be used by a small computing element to determine chess movements. Imagine a checkerboard that has optical fibers terminating under each square. When that square is covered by a checker, the light from the room will be blocked off and so no light passes. When the square is uncovered, by movement of the checker, then the light falls on that end of the optical fiber and so passes through it to its detector and so on.

There is no doubt but what other games might be devised to use the ambient light, or created light in beams or from bulbs, to cause activation of various functions. One such device being the "light gun" and target game. Optical fibers can connect the various target elements (or rings or such) to detectors so that as pulses of focused light fall on them, a panel could light up showing the score.

Sometimes it is convenient to have the light-detector (which converts light to electrical impulse or current) right at the point of light-entrance to the device, and sometimes it is not. When it is useful to have the computer element or coding element or decoding element some distance away, optical-fibers are an ideal to convey the light from the entree point to the device.

USING OPTICAL FIBERS IN MEASUREMENT OF ROTARY SPEED AND ANGULAR POSITION

NASA devised two optical sensors, one a 360 degree rotary encoder and the second, a tachometer to operate with a light source and a remotely located detector. The light source and the detectors were coupled to passive scanning heads through 3.65-meter fiber-optical cables. The devices were used on certain types of engines in which NASA was interested. This is another example of where the optical-fibers might be fruitfully used (in locations where the environment or conditions or situations preclude the possibility of having the sensors directly coupled to the detectors and where they must be located some distance away). Note also that it is implied that not only was the reflected light obtained through optical fibers, but the source light was also sent to the sensor through these optical cables. This type of two-way channeling of light rays presents unlimited possibilities.

CONSIDERATION OF OPTICAL FIBERS UNDER DYNAMIC EFFECTS

The attenuation of the optical fibers of both low and high (NA) single-mode fibers has been found insensitive to either low frequency dynamic tension or dynamic twisting in a study by ITT Electro-Optical Processes Division. The evaluation of the fibers with respect to dynamic bending showed that low NA (numerical operture) single-mode fibers were attenuated during the bending phase while the high (NA) fibers were unaffected. Coating materials having refractive indices either higher or lower than the fused quartz were applied to the single-mode fibers and were found to have no effect on the dynamic attenuation characteristics. No effects resulting from dynamic interactions were found in that all attenuation effects observed during the dynamic testing were explainable in terms of twists, tension, or bending.

USING OPTICAL FIBERS IN DANGEROUS SITUATIONS

There are some situations in which the use of electronic or electrical sensors could be very dangerous. One such situation is that of monitoring high-explosives or highly inflammable materials. Optical fibers with a light source presented to the sensor and the reflected light returned to a remote detector by means of optical fibers solves this kind of problem. The reliability of this type of sensing and the light transmission systems used is very high.

USE OF OPTICAL FIBERS IN COMPUTING ROLES

Sperry Research produced a report which described the development of a fail-safe optical data-bus system that uses several terminals. The data bus is in a serially-operated type of system. The system consists of a master terminal containing a CPU (Central Processing Unit) and LED sources and a photodiode detector and several remote send/receive types of terminals. The transmission system is composed of multi-mode monofibers and the data can be impressed on or tapped off at each terminal of the system. The purpose of the investigation was to develop a concept for the system such that it would continue to operate even if there was a power failure at one or more of the remote terminals.

After consideration of many types of terminals, an angle collimated terminal and a mirror terminal were constructed, both being compatible with Corning low-loss multimode optical-fibers with NA (numerical apertures) of about .15, both devices used about 85-micrometer thick crystals ($LiTaO_3$) to control the light flow in the optical fibers, using the electro-optic effect of these crystals. The mirror-terminal was finally selected for implementation to meet the research goals. It was proved that the system, finally consisting of three mirrors on terminals, when assembled into a complete system, met the objectives and proved the fail-safe nature of this type of system.

AIRBORNE USES OF OPTICAL FIBERS

There can never be any doubt but what a means to prevent the complex airborne electronics systems from becoming susceptible to electromagnetic interference (EMI) or electro magnetic pulses (EMP) which might be purposefully generated is important. It turns out that with the amount of computer activity used "on-board" most modern-day airplanes, that the miles of wire connecting the elements furnish ideal pick-up points for these stray and undesired radiations. Optical-fibers have no such susceptibility and their use makes the systems just that much more reliable and effective.

Airborne avionics systems are moving toward the uses of party-line multiplex data buses for the increasing number of digital signals used in modern aircraft. The opto-electronics developments now provide a data bus interface system but much work has had to be done to provide suitable couplers for the optical-fiber cables. It has been necessary to provide two-way transmission of the light-data into and out of the main data-buses. Each station on the bus must be able to communicate with every other station on the bus. In some cases a nine-port radial coupler has been used and pluggable optical interfaces have been employed with cable lengths of up to 30 feet. A maximum bus length in earlier trials was up to 100 feet and when the system used unipolar Manchester coding on the optical bus with a data rate of 10 megabits/second, a worst-case error rate of 1 part in one million bits was experienced.

But the critical area has been found to be in the couplings which permit two-way data transmission. Pulse compression techniques have been used to reduce the losses at such terminations and other techniques are being constantly devised. Some types of couplers under investigation are the so called "star" planar type couplers. Both theoretical and experimental characterization of 8 by 8 transmission couplers have been made with waveguide structures fabricated on a Nd-rich glass substrate using an ion-exchange process.

The structure patterns were formed on the substrate by means of a photolithography process.

Interfacing using switching techniques may employ an electro-optical switch which can couple any one of four waveguide channels. This may be selectively induced by the application of an electric field between intersecting strips of electrodes placed adjacent to the top surface of the slab and the other intersecting strips of electrodes placed adjacent to the bottom surface directly beneath the top surface. These electrode strips comprise primary electrodes to which constant voltages can be applied to induce optical paths partially through the slab. Also switching electrodes are used which are vertically placed and electrically isolated form the primary electrodes. To these are applied an alternating voltage that causes a proper and desired optical path through the slab to be accomplished.

USING THE FAST FOURIER TRANSFORM FOR PULSE ANALYSIS

The Fast Fourier Transform, adaptable to modern computers, can be used in analysis of the responses of optical fibers and their detectors. The National Bureau of Standards has an automatic pulse measurement system to measure the pulse responses of optical communications components and to compute their impulse and their frequency responses. For example, the measurement of the properties of two glass fibers (fiberoptics) and an avalanche photodiode using both a pulse (GaAs) laser wavelength of 0.9 micrometer, and a mode-locked NdYag laser wavelength of 1.06 micrometers has been accomplished. The measurements being performed in the time domain. In order to get the frequency domain data, the Fast Fourier Transforms were used with suitable computers. The impulse-type responses were obtained by deconvolution. This reminds us of the "early" days when we tried to evaluate the responses of some Hi-fi amplifiers we had developed and we used pulse inputs of various frequencies to test the responses. Square wave pulses were used of course.

STUDYING THE LOSSES IN OPTICAL-FIBER CONNECTIONS

McDonnell Douglas research department has been studying the optical interference losses between the transmitter-to-receiver interface systems. The pulse dispersion effects, rise-times of light transmitters and receivers, types of optical fibers available and detector sensitivity were all considered. Also many bussing systems such as the tee, star, and hybrid systems were carefully

148

analyzed. The matter of using a single fiber versus multi-fiber cables also was evaluated. They, of course, were interested in the use of optical-fibers on board a Space Shuttle. They considered a data bus on a space shuttle and then made an optical analog, using optical-fibers, which might replace the hard-wire system. System tests using a 9-port star data bus system for an optical-fiber bundle was considered as to losses, connector difficulties and efficiencies, data rate and such, and also the susceptibility of such a system to EMI (electromagnetic interference) in the range of 200 GHz to 10 GHz. Also, believe it or not, they tested such a system as to its vulnerability or nonvulnerability to lightning. It turned out that optical-fibers were idea in every instance and could (and perhaps should) be used on all future space shuttles. Thus, if that type of engineering is your ultimate goal, remember to seriously study the use of optical-fibers and all the elements associated with them!

THE JOSEPHSON JUNCTION

It is said that a Josephson chip can multiply an 8 bit by 12 bit array in just a couple of nanoseconds (10^{-9}), which is much faster than can be accomplished with currently conventional transistor-semiconductor circuits. Since speed and interference immunity are primary goals in future computer developments, this device becomes of primary importance. It is a switch arrangement, but such a terribly fast switch that some say it is 10 times or more faster than the fastest type switching circuits currently in use in integrated circuitry. The switch depends on a super conducting concept for its basic operation.

One concept of the switch, the Bell Labs Josephson-Atto-Weber Switch, named appropriately enough from the indicated acronym (JAWS) is shown in Fig. 4-17. In the off condition a current circulates in the triangular loop to the off line and ground. In the on state the loop junction resistance becomes infinite and, in effect, diverts the input current to the output line. So, the switch essentially acts like a short circuit element across a line. When the junction elements are at a low conducting state the current ignores them, and goes on to the output. When the junctions are in a highly conducting state, they short circuit the line and there is no output at the output terminal. To make the elements go from a low to a high conducting state, a control current must be applied, as shown. This development is in its most elementary state at present but it is so important that you can expect to be informed much about it in the future. It takes its place along with fiberoptics as a means to

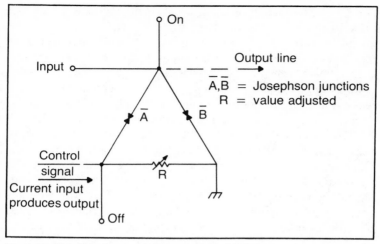

Fig. 4-17. The Josephson-switch (courteys Bell Labs).

increase the speed of operation of various devices such as computers and it will help to reduce their size.

One further idea needs to be expressed with respect to this switch. It needs to be very-very cold to operate correctly. So, in its tests, the Bell Labs put it into a Deward flask of liquid helium which keeps the *super-conducting junctions* at a proper (very cold) temperature. It is said that a circuit will be able to perform many operations and present them "in parallel" in its output when many of these switches are formed into an array. *Cryogenics* is a name which has been applied to operating things in a super-cold environment.

SOME RANDOM OPTICAL CIRCUITS

There is always a desire when reading about or studying new concepts such as optical fibers and the circuitry associated with

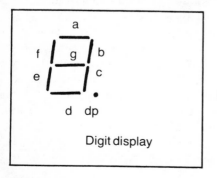

Fig. 4-18. A seven-section LED.

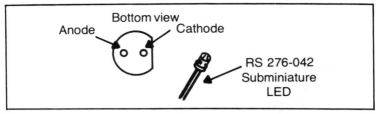

Fig. 4-19. A LED and connections.

them to want to get some "hands-on" experience. To this end we conclude this chapter with some random circuits that use LEDs and/or other optics in their operation. One of the most common uses for the light-emitting diode is that of displaying numbers. One such unit for doing this is the Radio Shack 276-060 LED 9-digit display block. It uses GaAsP LEDs for display and it can easily be interfaced with TTL DTL, or MOS units. It is illustrated in Fig. 4-18. Of course these type displays can be obtained in everything from a single number unit to a multi-unit.

Now we need to examine the connections for some LEDs. The first in Fig. 4-19 is a subminiature red LED lamp-type with a diffused lens. You apply a small voltage (1.6 volts) and it glows. In Fig. 4-20 we have a single LED of the jumbo size. It uses 1.75 volts for its operation. There are many other types of LEDs that might be just what you are looking for. Ask about other types at your radio parts store.

Sometimes you might want to make a flasher unit with a reasonably high rate of "flashing." One such circuit is shown in Fig. 4-21. In Fig. 4-22 we find a receiving circuit which can produce an output when a light falls on the input diode. Of course all these circuits may have to be experimented with an value of parameters (resistors and such) adjusted to get the kind of operation you want.

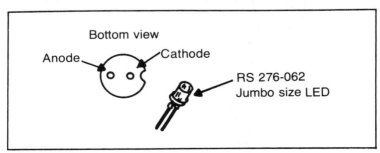

Fig. 4-20. A jumbo LED and connections.

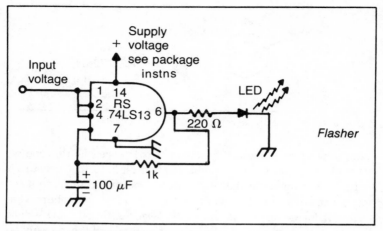

Fig. 4-21. A simple LED flasher circuit.

We recall seeing a nice friendly robot, Fubar, (Fig. 4-23) in which the enterprising company had placed some flashing LEDs in his "mouth" area so that when he "spoke" (actually he was radio-controlled and so his robotmaster probably spoke for him) his mouth flashed in various patterns which made it look for all the world like his lips were moving. Quite clever! Also, flashing LEDs, (probably in circuitry like that of Fig. 4-21) came on when the robot was activated. These were visible in some of the body areas (mouth-head-chest). So, if you are "into robotics" you might consider the

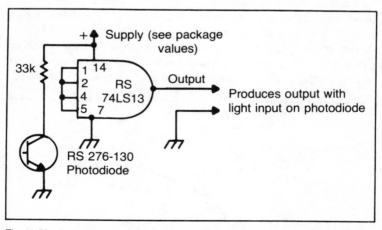

Fig. 4-22. A circuit that produces an output voltage when light falls on the photodiode.

Fig. 4-23. Fubar the robot uses LEDs to create special effects.

use of these types of elements as a means to enhance the mystery and fascination of your mechanical (but loveable) monster.

We have gathered together some circuits that we found very fascinating both for their possibilities and use in some exotic applications, and for their illustrative indications of how opto-electronic

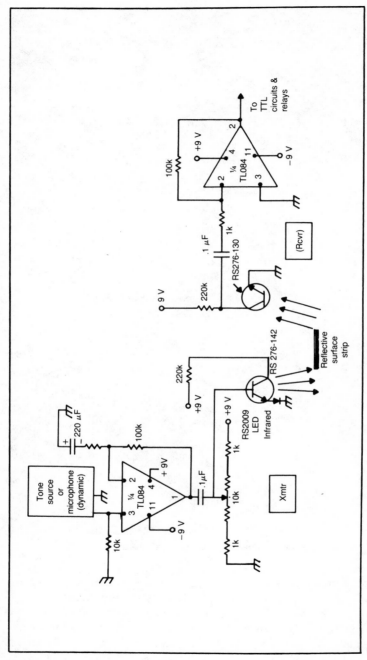

Fig. 4-24. One type of circuit for robotic path control, for an alarm system, or for a communications system using infrared light rays.

circuits are put together. In Fig. 4-24 a simple infrared transmitter and receiver are shown which might be used in a path-finding application of a robot. We caution that you might have to do some experimenting with the resistor values. When you purchase the parts (say from such supply houses as Radio Shack or a similar source) they will no doubt have some circuit diagrams as part of the package. These might also be used effectively.

One LED transmitter or flasher can be built around a 555 integrated circuit as shown in Fig. 4-25. Also shown is a LM 3909 type unit in a very simple connection to produce some light flashes. A pulsing infrared light might be used to give a robot some idea of direction—that is, it might be made to "home" on the flashes. Or it might be used to warn the machine of possible confinement areas or obstacles or trouble spots which it should avoid. Control of the flashes can be used to send various control codes by infrared light. We leave that idea for your development.

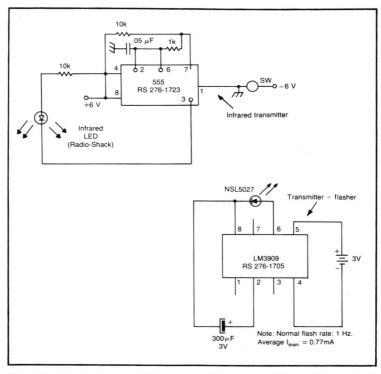

Fig. 4-25. Circuits showing LEDs as transmitters or generators of infrared light rays.

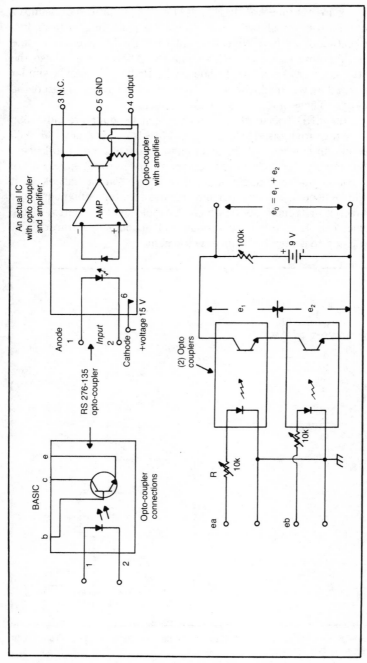

Fig. 4-26. Some opto-couplers and integrated circuits.

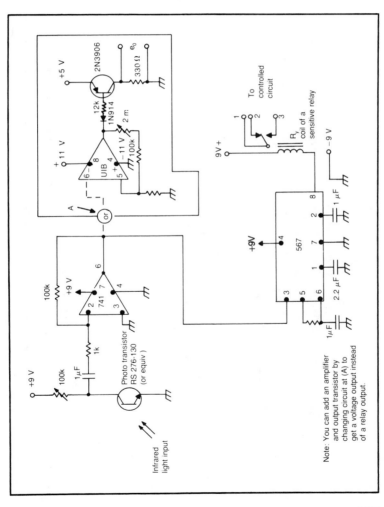

Fig. 4-27. A receiver for light-control of another circuit.

157

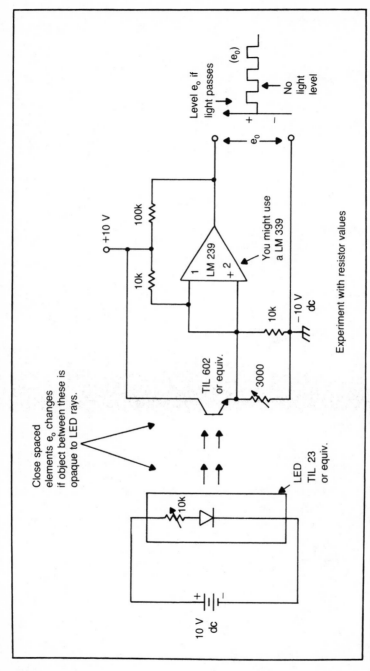

Fig. 4-28. A tape reader using a light source and detector.

The opto-coupler is an ideal way to isolate one sensitive circuit from another (power control circuit from an input circuit). In Fig. 4-26 we see the basics of such circuitry and some integrated circuit schematics which may be useful to you. There are many types of opto-couplers. Some use triacs for control. If you want to make your own coupler, there is no reason why you cannot use an LED type transmitter coupled into an optical fiber and connect a receiver photodiode to the other end to convert the light to current-voltage. It is interesting that two opto-couplers can be used to sum some input voltages as shown. We often call this "mixing." It can have application when you need to have two sources (sensors) provide some inputs and then not have a reaction until *both* sensors say it is time to do so. Or, until one sensor has such a large output that you need to have some action taken by the (robotic) machine.

We are always interested in receiver circuits for light rays or light pulses. One such is shown in Fig. 4-27. In this circuit a sensitive relay is caused to close when light falls on the sensor (photodiode or phototransistor) and hold closed as long as the light is present. There is an alternative circuit arrangement using the U1B amplifier or equivalent that feeds a single transistor output element, and this provides a voltage to activate something else (perhaps another amplifier). In any event, these are two possibilities for your consideration and use.

Finally, we show how you might connect some circuits to provide a continuous light source that falls on a phototransistor receiver to produce a signal which can be amplified and presented at the output of the LM 239 amplifier (Fig. 4-28). If you block off the light path, you might get no signal output. When the light passes to the transistor detector you get a full voltage output. This type of circuit might be a tape-reading machine control unit for a numerically controlled machine (see my book Advanced Robotics TAB book Number 1421). Notice that other types of IC's might be used for the amplifier. Don't be afraid to experiment. Who knows, you might come up with a "better mousetrap" and become rich and famous!

Chapter 5
The Importance of Lenses

We have learned about refraction, reflection and the basic composition of light. We know that when such rays pass through mediums with various indices of refraction they can be bent or refracted. We also know that if we have a light wave front hit upon a power type of surface (a mirror comes to mind) those light rays are reflected from that surface at specific angles. We think of the old rule: "The angle of reflection equals the angles of incidence." Now we are to be concerned with an examination of the concept of focusing of light rays through various types of lenses. We need to know about lenses because they will be used with light sources which, in turn, pass the light rays into the optical fibers we have been discussing.

THE OPTICAL FIBER CABLE OR MULTIPLE FIBERS

For just a moment we want to consider, again, the diameter of an optical fiber. We know it is very small, perhaps just about the size of a human hair in many cases. Of course, it could be made larger depending upon the type of glass material used in the core. The cladding causes the refraction of the light rays inside the core, and makes the light bounce around and "go down the core" from the input end to the output end. It also is very thin and so even if a single fiber is finally covered with a "jacket" of plastic to protect it and give it strength, it is still not a very large diameter item.

So we consider the idea of a "bundle" of fibers (many of them are parallel to each other) even though they may be twisted in total length. Intuitively we gain the "feeling" that this could provide some strength to such a light cable, might help to avoid problems with the breakage or imperfections of one strand, and provide enough diameter to be accurately focused into or out of using lenses. It is possible to have LEDs and lasers of special types which might be small enough to become the end parts of a single optical fiber strand. These could have the necessary lenses incorporated into their structures, but it is a little difficult to imagine, especially for those of us not actively engaged in experimental research and development in this area of light transmission.

To avoid complications and worry about that smallness, while we are considering the concept of lenses and use of lenses with optical fibers, let us consider that either we have an optical fiber of some diameter into which and out of which we might be able to focus light rays, or that we have a "bundle" of similar type fibers, each conveying light rays to the end-point or surface. The effect at the input and output is that of a single larger diameter fiber. Now we can consider the use of lenses a little more and we are a little more at ease with the actual physical geometry of the situation.

SOME TYPES OF LENSES AND FOCUSING POSSIBILITIES

Let us consider the spherical concave mirror first. We know that this type device will reflect the light rays if the basic principles of geometric optics (having the surface large when compared with the wavelength of the light reflected) are not violated. Recall that we are considering light in the micron (10^{-6} meter) millimicron (10^{-9}) or angstrom (10^{-10}) region. Visible light being from about 450 to 675 millimicrons. Infrared light being higher (shorter wavelengths). Examine Fig. 5-1 which gives the concept of a spherical concave mirror. There can be problems with this type reflector. If the source and image are not precisely located for a given reflection just a certain part of the mirror, there will be no reflection of the source seen at the reflection designated position.

It is important to us here to understand that a mirror will reflect light only from the "front." If there is light behind the mirror, it will not normally pass through, even though in theory there is a consideration of the path of rays as though they do pass through the mirror. Later on, when we discuss lenses we shall be talking about such things as a real image and virtual image.

If we assume that the light rays come from a long distance away so that they are parallel, or that the mirror is small or equivalent in size to the source which might be the end of an optical fiber bundle

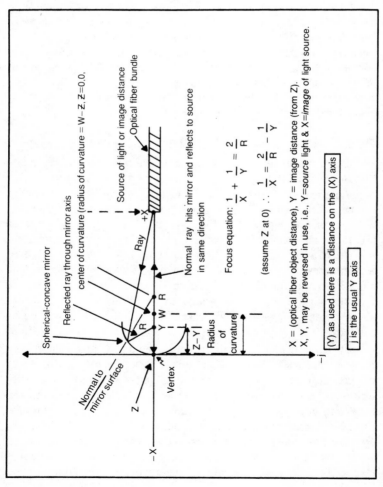

Fig. 5-1. Details of a spherical-concave mirror. (Note that "y" specified here is actually a distance along the "x" axis.)

(polished flat and perpendicular to the direction of the emitted light rays) then we find that the image point (real or virtual) is called the *focal point* (F) of the mirror. If we assume that the light rays come from a great distance (so that they can be assumed parallel) and that the source or mirror is at the (F) focal point and is fixed so that it does not move, then we can assume the rays will impinge upon the mirror and emanate from that source.

Due to the curvature of a concave or convex mirror, rays might be considered to converge or diverge. If they converge, we have a concentration of the light rays. We find that a simple statement of geometry: [image (i) = ½ r (radius of curvature) = F (focal distance)] tells us where the focal point of such a mirror should be located. It is at ½ the radius of curvature distance from the *vertex*. The classic equation expression the focal distance is:

$$\frac{1}{O} + \frac{1}{i} = \frac{1}{F}$$

O = source distance (object distance)

i = image distance

F = focal length (positive if image is real, negative if image is virtual)

PLANE MIRRORS

Light which impacts on a plane mirror is reflected from that mirror through the angle of incidence. Or, to put it another way, the angle of reflection is equal to the angle of incidence. This is easy to check out in a simple experiment. Figure 5-2 might help the visualization. The plane mirror is important in some types of lasers. In a ruby rod laser the stimulated light energy bounces back and forth until it emits from one end of the rod. The other end is a polished opaque, plane-mirror surface.

When you view the image from a plane mirror be careful to observe the interchange of "left" and "right." If you think about it for a moment you'll realize that the man you shave in the mornings shaves with his left hand on the left side of his body if he is right handed and vice versa. A woman will apply make-up with the opposite hand, considering the body image as reference, than she actually uses in the application of such makeup. Our minds adjust to this phenomena and we usually aren't even aware of it. But think about it next time you face yourself in the plane mirror. Remember a plane mirror reverses right and left and this may or may not be of consequence in our reflection of light rays and light energy.

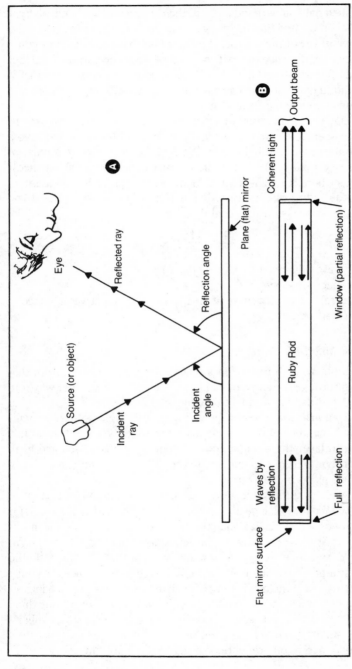

Fig. 5-2. Reflections from a plane mirror. A flat mirror plate reflects light images (A). A ruby-rod in a laser must reflect light to stimulate emission of laser beam (B).

THE FOCUSING PROBLEM

When we consider getting light into an optical fiber (singly or a bunch of fibers all terminated in a flat and polished end) and we consider using a reflecting device such as a mirror or use a focusing device such as a lens, we must also consider the angle of entry into the fibers of the light energy. If the light energy is presented through too wide an angle the angle of reflection from the cladding may exceed the "critical" or Brewster angle and the energy might not propagate through the fiber. We want to concentrate the energy and light rays within a cone which will have the requisite acceptance angle for the mode of propagation we desire. Size becomes important (the physical size of the lens or reflector and the relative physical size of the optical fiber cable or strand). If the geometry of the reflector or the lens can be made compatible with requirements for reflections and focusing and is small enough so that the cone angle is within tolerance, then "we have it made." But, sometimes that takes some "doing." The small size (diameter) of the optical fibers makes fabrication of such lenses and reflectors somewhat difficult. Credit is due our wonderful manufacturers and their research departments that they have been able to accomplish such a formidable task! See Fig. 5-3.

We readily see that if the mirror is small and of a size close to that of the fiber, it not only loses much of its ability to gather light rays, but it can present what it gets within a relatively small cone angle to the end of the optical fiber. The larger mirror gathers much light from a source but the angle to the fiber may be much larger. Ideally, a source would have a lens about the same size as the fiber.

FORMING LENSES IN ENDS OF OPTICAL FIBERS

When we consider that the focusing effect of a lens is due to different indices of refraction between air and the material of the lens, then it is natural that we wonder if it wouldn't be possible to form a lens out of the optical-fiber material itself. It would actually make no difference whether the end of the fiber were the end of a bundle of fibers or a single fiber. If there are many fibers in a bundle we might assume that various light rays would go through one fiber and other rays would go through other fibers.

Figure 5-4 shows a source of light energy at O and the two rays, one going directly into the glass-material through the radius of curvature of the end (r) terminating as shown at (I) meeting with the refracted ray which hits the curved surface and then is refracted to

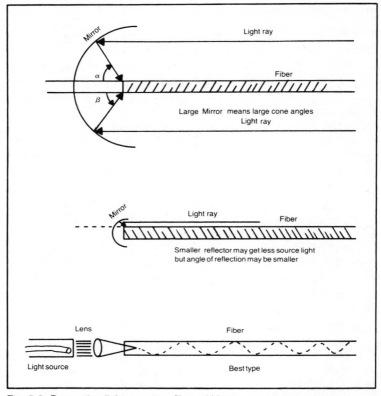

Fig. 5-3. Presenting light rays to a fiber within an acceptance cone angle.

also meet at (I). The fact that the two meet at (I) in a lens effect is important in lens considerations. We have shown that the two rays are now inside the optical fiber. They arrive at (I) at such an angle that they can propagate down the length of the fiber(s) to the opposite end. Thus it would seem from this diagram that we capture more of the light energy in this manner.

It is to be understood that it is possible to form the end of the optical fibers into such a type lens. By heating to the proper temperature, the glass will become plastic and can form or be formed into such a lens. Some skill is required and very careful temperature control is needed so the glass won't "run" or "ball" or otherwise form something we don't want. But it can and has been done, and this could be important in coupling applications and capture applications.

Next we assume that the reverse kind of situation might be

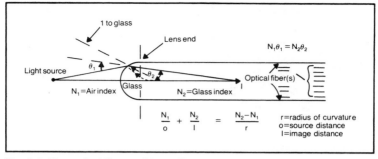

Fig. 5-4. The optical-fiber end-lens effect.

possible if we make a lens out of the terminal end of the fibers. Physics shows us that with proper refractive qualities of the fibers and medium beyond the end of the fiber(s) that a refraction that is the reverse of the input effect shown in Fig. 5-4 can take place. The source becomes the focal point and the I point of that figure must be a cross-over of the rays inside the glass material so that this becomes an "inside source" origin of the rays.

In a manner of thinking we have the "thin-lens" effect at the end of the optical fiber. One curved side of the thin lens is not really formed but is present as illustrated in Fig. 5-5.

There are other types of lenses and effects that one might presume can be obtained using materials with different indices of refraction. What we find important here is that focusing and small source rays are requirements in optical-fiber systems, even when the fibers are in bundles. This means that we cannot handle optical

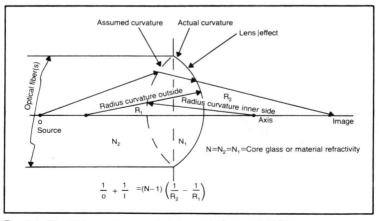

Fig. 5-5. The "thin lens" concept at the end of an optical fiber.

fibers or optical-fiber cables like we would handle hard wires. The end connections, joining elements, coupling elements, and feeding and receiving elements will have to be carefully designed. Lens effects must always be considered. You just don't clamp the ends of two fibers together as you might join a pair of wires and expect the transmission of the energy to be equivalent. Much more care and caution is necessary in the handling of the optical fibers or optical-fiber cables.

HOW MANY CHANNELS CAN THERE BE IN A SEGMENT OF THE LIGHT SPECTRUM?

One of the most important concepts related to the use of light as a transmission medium and the use of optical fibers to guide these light rays is that of gaining more channels. We "mix" data and TV and radio and telephonic communication on wires of the "two pair" variety and the "coaxial" variety, and we find that we have not enough "room" or channels to carry all the information required. More channels are needed and will continue to be needed as our need for data transmission and information increases. How to get the extra channels is the question.

Let us now consider a normal coaxial cable transmission system that can transmit frequencies up to several gigahertz. The band-width or channel-width for good telephonic communications is about 3,000 hertz/second (3×10^3). Radio broadcast channels require about 10 kilohertz for their channel spacing. In the microwave region channels are some 250 to 300 kilohertz in width. If we want to find the number of channels available in a given span of frequencies, we simply take the width of the channel to be used, plus its separation channel, and divide this into the width of the space available. The answer is the number of channels that can "fit" into that spacing. An example: if a two wire system can transmit frequencies from dc (direct current) up to 1 MHz (one million hertz/second) and good telephonic communication requires 3,000 hertz channel width, then:

$$\text{Channels} = \frac{0 \text{ to } 10^6}{3 \times 10^3} = 333.333333...$$

A letter over 333 channels per hard wire pair! Obviously these aren't enough channels to be practical, so consider a coaxial cable which can handle, say up to 1,000 megahertz/second. Now our figures become:

$$\text{Channels} = \frac{1000 \times 10^6}{3 \times 10^3} = 333{,}333.33 \ldots \times 10^3$$

or several thousand times more than the wire pair. The line is also probably smaller even though terminal receiving and sending equipment and couplings may be more complex.

Notice that in these computations we are assuming a *simultaneous* transmission of all channels. Actually, time-division multiplexing or frequency-division multiplexing increases the traffic handling capability. So we are "up-to" the consideration of optical fibers with their wide band capability. Let us assume for this imaginary example that we can transmit visible light in the region of 450 to 650 angstroms over such a cable. It is to be realized that an optical fiber here might be used singly and it is, indeed, very small in diameter! So we are considering a visible light band of 200 angstroms, which represent a space available channel width of:

$$\text{Hertz} = \frac{2.997924 \times 10^8}{200 \times 10^{-10}} = 14{,}989.62 \times 10^{18}$$

For our communications problem requiring a 3×10^3 bandwidth we find:

$$\text{Channels} = \frac{14{,}989.62 \times 10^{18}}{3 \times 10^3} = 4996.54 \times 10^{15}$$

If you'll write 15 zeros after that number you'll see why fiberoptics is important, at least in the telephonic sense!

SOME PROBLEMS WITH CHANNELING USING OPTICAL FIBERS

When we discuss the number of channels that might be obtained over optical fibers using light rays, we do not at first consider the "other effects" which tend to limit these channels. We need to at least be aware of what they are. In the first place we do not have exactly pure light. That is, it isn't exactly on a single line of the spectrum. Lasers come closer to having this single line spectrum than any other light source, but one can expect some "unexpected effects" which might tend to broaden the light spectrum even from a laser source. Then it encompasses more frequencies (instead of a single one) and we have a narrower channel spectrum to fill because the non-usable channel itself is wider. Fewer channels can be accommodated in a given frequency range if the nonusable portion of the source channel is wider.

Then there are other effects, which include intermodulation distortion (the by-product frequencies of input frequencies or pulse rates) and intermodal dispersions of frequencies called by the non-purity of the index of refraction of the optical-fiber line or cable throughout its length. We have mentioned that pulses might be distorted if a line has much intermodal dispersion. This causes the pulses to be so rounded and delayed in transmission that problems exist in recovering the intelligence from them. So the channel spectrum required for good intelligent transmission may not be calculated as simply as we have indicated in our elementary example.

But there will be so many more available channels, that even with some of these problems, optical fibers still seems to offer the *best* current solution to the transmission of intelligence and data in our ever increasing complexity of interconnecting requirements between and among various "senders" and "users."

Just think for a moment of the channeling in another way. We have assumed an analog type of frequency, a continuous wave type of tone, voice, or data pattern. But consider the effect of pulses and pulse rates. To transmit a pulse requires a wide bandwidth, some say it must be: bandwidth equals $\dfrac{1}{\text{pulse duration}}$ to use an old radar formula. This might be the minimum baud width required to recover a reasonably shaped pulse. Then we add to this the pulse rate itself. Assume that the pulses are transmitted at half a million hertz/second. That represents another frequency component into the line and its harmonics. If we assume a varying pulse rate (depending upon the type of intelligence represented) we then have a multiple of equivalent frequencies present and their products and sums and differences and so on. One can think of a system in which "time division" multiplexing will restrict the pulse rates to only one during an "interval of time" being present in the line so the multiple by-products are not problems. That, of course, is a possibility.

SOME PATENTS CONCERNING OPTICAL FIBERS

It is always interesting to examine patent information about optical fibers. As time goes on there will be more and more of them as the use of these fascinating elements increases. One patent is directed to a security device that detects tampering with a secured inclosure. A fiber-optic *bundle* is looped through a closure and secured at opposite ends of the bundle to a snap-together connector.

An intermediate length of the fiber-optic bundle surrounds the snap-together connector preventing access to its locking mechanism unless the fibers are cut. After installation, light is passed through the fiberoptics and a particular pattern is generated at the viewing end of the connector. Tampering with the inclosure will cause the individual fiberoptics to be disturbed so that subsequent viewing of the particular pattern will indicate something different than that which would be seen if no tampering were done. Thus one knows that the lock has been tampered with.

This brings to mind a possibility which might come about one of these days and that is a window glass might be made so that an optical fiber strand curls back and forth throughout its area. This would not be visible ordinarily, nor would it impede the normal transmission of light through the glass when viewed from the front or rear. But an invisible light pattern could be sent through this "channel" and if the glass were broken in any way, the light-to-electronics receiver would then sound an alarm, much like we now get with hard-wire type conductors around the windows, or the "vibration sensing bugs" on the windows.

There has been invented a three dimensional memory having a vastly increased storage capacity, using optical-fibers. The memory block has a matrix of cylindrical cavities, each of which has a fiber-optic light-guide embedded inside. Each light-guide is composed of a cylindrical core having a first index of refraction and a cladding surrounding the core that has a second index of refraction. This second index of refraction is smaller than the first. There are a number of spaced deformations formed at the cladding-core interface that permit the light to leak out laterally. It is presumed that this light contains the memorized elements.

Another interesting area is that of the use of optical fibers in a hydraulic diagnostic monitoring system. This system is to warn of failures in the hydraulic system components by using on-board sensors that continuously monitor various failure indicating parameters. The monitoring system consists of three basic types of sensors; analog, discrete, and fiberoptics. These sensors feed information to a self-contained, centrally located display panel through the various interface circuits that are easily replaced and readily accessible to maintenance personnel. Also monitored are the display panel indicators that might fail and thus indicate a malfunction when there actually is no malfunction.

We note an invention that is a remotely controlled solid-state relay with a fiber-optic input. The relay is powered by the circuit being controlled and thus uses no power of its own. A trigger circuit powered from the ac line is actuated by back-biased PIN diodes. The trigger circuit controls a switch that is a pair of silicon-controlled rectifiers in series with the ac source and the load. One presumes that the light input through the optical fiber triggers the circuit, which, in turn, permits the rectifiers to function and thus supply power to the controlled device or circuit.

DEVELOPMENT OF FIBER-OPTIC DEVICES

One report concerned with coherent fiber-optic testing techniques fits well into our discussion. Of course, we know that any development has to be proved. So it is worth considering the following from *the symposium of Photographic Instrumentation Engineers*. "We have found that reliance on theoretical models or incomplete manufacturing data is not adequate for predicting the results of combining optical and electro-optical components into systems. The complex imaging characteristics of each component must be accurately known if reliable predictions are to be expected. In-house testing is desirable when possible. Some components have well established evaluation procedures *and others such as coherent fiberoptics* (as of 1979) requires test techniques, some of which are well established and others which must be devised, to find out if suitability is accomplished and if complete characterization (or definition of operational characteristics is to be found). This is especially true where changes in magnification are a feature of the bundle (of fiberoptics). The technique used for testing can include image rotation, magnification, image size variation, distortion, shear, transmission efficiency, transmission variations, transmission defects, flatness and acceptance angle. These are described in the symposium's paper.

When we consider the types of tests as specified here, we begin to realize what it means to insure that, say, a given batch of optical fibers will be usable in some particular application. Continuous testing of this order of magnitude may then be a mandatory requirement in a manufacturing plant. Also the use of new and improved testing devices and methods will be of constant interest in this fast growing field of science and engineering.

Some definitions of devices used with optical fibers could be of value. An optical *"access"* coupler and a *"duplex"* coupler have some special meanings. Access couplers are for use with multi-terminal communications systems. The *duplex* coupler is a device which allows a single fiber to be used for data transmissions in two directions. Such devices may consist of special grooves of different depths and widths that have been etched into silicon along natural planes so that one may position fibers of different diameters properly, and securely in the grooves, and also to provide in couplers, reflective silvered surfaces for proper reflections. Optical communications signals pass through smaller diameter transmission lines to a larger line through this kind of device. The light going from the smaller line to the larger line may be reflected by these reflective surfaces, and varying them permits adding the intelligence desired.

In some coupler applications it is necessary to couple energy that comes from a light-type waveguide into a more ordinary rf waveguide. This can be accomplished with a single fiber held in a capillary tube and positioned by a micro-positioner so that the greatest light output is obtained. The fiber is then secured to the waveguide by means of epoxy at one end of the waveguide channel. The two elements are then positioned in the capillary tube in such a way that they cannot be rotated.

It is to be realized that reflective surfaces can be used to "channel" and "guide" the light energy from one end of one optical-fiber cable into another cable. The positioning of these reflective surfaces must be exact and they must be small as we have previously indicated. They must be rigidly positioned along the fiber ends such that no movement or distortion of position is possible. But, considering such an arrangement, we can visualize how a multiple of combinations of light energy might be summed together in such an arrangement. Also we can visualize how an input line might convey the light energy to a given reflective position and then that energy be subdivided through the reflective position so that many individual strands can be energized with the incoming rays.

While considering the use of optical fibers in transmission systems, we become concerned with the mechanical strength of the fibers. The "cable" must be handled in the field and so it must have some strength. In one effort to increase the strength of these optical fibers a ceramic coating is applied over the optical fiber core. It is an inorganic material impervious to moisture and chemically corrosive materials. Presumably it is not the cladding material but is applied over that type material in the "drawing" process.

SHOCK WAVE EFFECTS IN OPTICAL FIBERS

Although the optical fiber is impervious to electromagnetic

effects and can be used in many applications where large magnetic fields exist, it has been found in some applications that shock waves in solid materials can turn the fiber opaque and prevent its transmission of light rays. Thus, if the fiber may be subjected to these shock waves, one has to take precautions to protect them. There are some beneficial effects from this phenomena as pointed out in some studies. Due to the fact that the shock wave does make the fiber opaque, it can be used as a sensing and measuring device to generate a signal when the shock wave is present and absent. In some research applications this timing capability can be of importance. Of course we are aware that other timing applications can generate from this "light-no-light" transmission concept. This can be even to the point of severance of the fiber due to some mechanical or chemical or other action that interrupts the light transmission at a specific time.

If the fiber is accurately positioned with respect to a receiver so that the light rays emanating create a signal in the receiver, if anything disturbs the alignment of the fiber and the receiver, the signal will vary or cease at that instant. Thus a marker of the time of the event is generated.

A SHORT REFRESHER COURSE IN OPTICS AND LENSES

What is light? Light is a radiant energy which affects the eye and enables us to see. Light can be visible to us and can be invisible to us. Light moves at a specific and unvarying speed. Light rays are said to be parallel rays. But light is said to be a component of the electromagnetic spectrum, and James Clerk Maxwell proved this to be the case. This is important because we can then consider that light is simply the "far end" of the radio-radar-microwave spectrum. A chart such as shown in Fig. 5-6 is helpful to visualize this concept.

The spectrum is logarithmic, of course. And, as we have already indicated to some extent, the visible portion of the spectrum is from about 450×10^{-9} to about 650×10^{-9} meters. The word millimicron is often used in optics with the abbreviation $(m\mu)$. It is interesting that the lower frequency of our sight is the violet-blue region and the upper frequency is in the red region. The eye is most sensitive in the green-yellow region. Now you know why lots of signs are colored yellow with black, or dark lettering. Traffic lights are yellow and green and (wouldn't you know it?) red on stop!

Of course, all people do not see the same colors with the same vividness. But, we won't worry about that here. Our optics systems, being electronic and electromagnetic can detect whatever

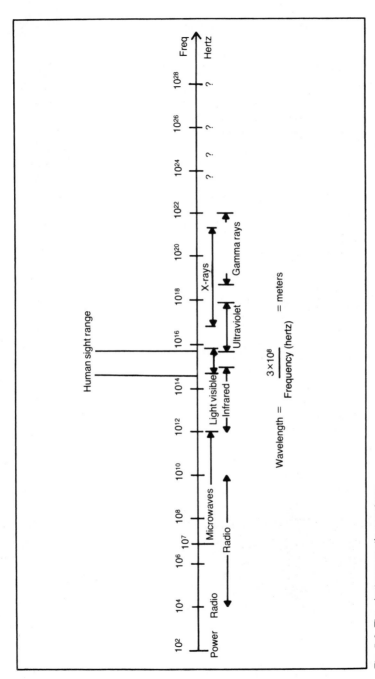

Fig. 5-6. The electromagnetic spectrum.

175

range of frequencies (colors) we design them to detect. Notice, however, that our optics systems may be somewhat "color blind" if we don't properly design them. This means they won't transmit the light rays of some colors (frequencies) as easily as they can transmit other colors and frequencies unless we design the material properly. It is said that the limits of human sight are from about 4300 angstroms to bout 7000 angstroms. That word means 10^{-10} meter.

Computers and other electronic devices can be used to "enhance" the things we look at (pictures or TV screens etc.) so that we can see colors which we might not ordinarily see. Note this on the next weather-news broadcast you see on TV. Most such programs use color enhancement to show wet-dry areas. Fiberoptics requires that we generate certain colors and frequencies of light for proper transmission through the fiber-optic channels. You can imagine some of the problems if that light color is not constant or at least relatively so. Light sources are very important to us. Some LEDs do not generate constant frequency outputs. Some lasers require feedback and automatic adjustment of electrical energy to maintain a relatively constant power and frequency output. Since the frequency is on a logarithmic scale, a very small change will result in a considerable effect. Much effort is needed to maintain the constant frequency of the light source.

WHITE LIGHT

What we perceive to be "white light" is not a single wavelength of light at all. Sunlight is considered to be "white light" and we know it is composed of all wavelengths of light energy. We also know that we can artificially mix various wavelengths of light energy to form what the eye will see as "white light." In a way, this is like "white noise" in electronics systems. The noise is composed of all frequencies of noise mixed together.

SOME CONSIDERATIONS OF LIGHT ENERGY

If we consider light rays to be simply an extension of the electromagnetic spectrum, we are in agreement with most modern physics texts. Under this consideration, the energy in the light rays is described by the familiar *Poynting Vector* of a radiated wave:

$$S = \frac{1}{\mu_0} \overline{E} \times \overline{B}$$

This is illustrated in Fig. 5-7. The resultant (S) vector, is the cross-product of the electric vector (E) and the magnetic vector (B). This shows the direction of wave motion and the vector cross-point mathematically gives the magnitude of this wave.

But, what has worried some physicists is that the light rays also exhibit a momentum (pressure) effect. We know this because we can measure the pressure on some types of surfaces—recall the small four-leafed axle in a vacuum bottle with its silvered leaves? When placed in the sunlight so the leaves reflect the sun's rays the axle will spin in a direction to cause the leaves to revolve away from the sun's rays. The speed of rotation is dependent upon the brightness of the sun's rays. In the glass-vacuum container, there is no air friction—or at least it is minimized—which otherwise would slow down the rotation. In modern space-platforms this pressure-effect of sunlight may have considerable importance. Some studies have actually considered a "sun-sail" which might be extended from the body to gain pressure to cause motion in space. Fascinating idea isn't it?

It has been determined by experiment (after Clerk Maxwell said the effect would be evident) that the momentum given to a body could be calculated by some equations that are very simple in expression. (We are always amazed at how most of the world's important equations are simple three letter combinations such as $E = Mc^2$, $E = IR$, $F = Ma$ etc.) Here, again we find the same type of expression:

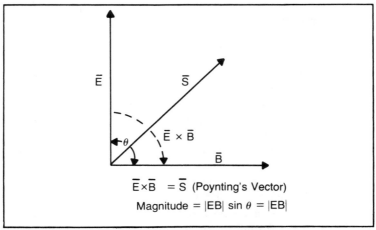

$$\bar{E} \times \bar{B} = \bar{S} \text{ (Poynting's Vector)}$$
$$\text{Magnitude} = |EB| \sin \theta = |EB|$$

Fig. 5-7. The Vector cross-product gives Poynting's vector.

Momentum when ray is totally absorbed by body: $\dfrac{E}{c} = m = p$

where E is light energy
c is speed of light
m is the momentum also written p

Momentum when energy is totally reflected: $m = \dfrac{2E}{c} = p$

As you'd expect, in practical applications where energy is somewhat absorbed and somewhat reflected, the value would lie somewhere between these values. These equations are important in calculating, for example, the size "sail" needed to cause a space vehicle to move at some value of acceleration through space, assuming no air friction magnetic friction or viscosity or such. Since acceleration is given by:

$$a = \frac{F}{m} = \frac{\text{Force}}{\text{mass}}$$

then, if we can calculate the force produced by the sun's rays on a given size "sail", and if we know the space ship's mass we can determine the acceleration of the body through space. Recall that a constant acceleration, no matter how small it may be, can result, *ultimately* in *almost* the speed of light! As we know we cannot reach the speed of light because Einstein showed that when a body reaches this speed its mass becomes infinite.

THE QUANTUM THEORY AND THE PHOTOELECTRIC CONCEPT

It used to be that because some substances emitted electrons (called photo-electrons or photons) the electromagnetic theory wasn't considered usable to explain this effect. The *quantum theory* could be used for such explanations. There were other effects about this photon emission which furrowed brows (the velocity of emission is not influenced by the intensity of the light rays) the emissions increase with the frequency of the light radiations. The amount of emission does vary with intensity of radiation impinging on the target.

So, how was all this explained? It was explained by assuming that the light energy consisted of small energy packets called *quanta*, or photons, and these had different energy values, depending on the light frequency or wavelength. Again, that simple three-letter type of equation described the energy content:

$E = hf$ (E = energy, h Plank's constant, f = light frequency in Hertz) and so the energy related to the electrons in order to satisfy Einstein's equation became:

$$\tfrac{1}{2}\,mv^2 = hf - w = hf - hf$$

m = mass
v = velocity
w = work function or energy required to remove one electron.
hf_0 = lowest value of hertz at which emission takes place.
hf = frequency of light applied

THE PHOTOELECTRIC EFFECT

Of course, the photoelectric effect is what interests us mostly here. This effect is that if we impinge a certain type of light on a certain kind of substance under certain conditions, the substance will give-off electrons of a type that we can measure as an electric current. Isn't that what we want in our optical fiber transmitter-receiver systems? Millikan won a Nobel prize for his work in connection with this effect in 1923. It was found that this photoelectric effect was a kind of "surface phenomena" and that if the surface was contaminated with grease, dirt, oxide films, or other such inhibitors, then the photoelectric effect would be vastly reduced or nonexistent. This is very important in the manufacture of photoelectric devices we use as receivers and converters of light rays to electric currents.

There are some contradictions when considering the "wave-theory" of Maxwell to explain light wave radiations. The quantum theory seems to fill the gaps and permits a more unified and acceptable explanations of these contradictions. Some of these are: Wave theory says kinetic energy of photoelectrons should increase with an increase in light intensity. Experiment shows this not to be so. The kinetic energy remains independent of the light intensity. Wave theory also says there should be a photo-effect no matter what the frequency of the light rays may be. This is not found to be true. There is a cut-off frequency which is the lower limit of the effect. There should be some delay in the emission of photoelectrons when the light rays may be. This is not found to be true. There is a cut-off frequency which is the lower limit of the effect. There should be some delay in the emission of photoelectrons when the light rays are weak. This is not so. There is no delay no matter how weak the rays may be.

Max Planck believed that light traveled through space as an

electromagnetic wave, but Einstein showed that light had to travel as small particles. Einstein's expression (proved in experiments by Millikan) seem to be most valid *when considering the photoelectric effect*.

What we find in this analysis of the use of optical fibers and light is that we use the wave theory to analyze the propagation of the light rays through the fiber, but we would then consider the quantum theory when analyzing the devices which convert those light rays back into electrical currents for use by computers, and other such types of equipment. You might want to investigate these theories in much more detail than we have presented here. Just remember that, currently, scientists consider light to have a dual nature (behaving like a wave in some circumstances and like a particle in other circumstances). The particles are called photons.

ON THE DUALISM OF LIGHT RAYS

Louis deBroglie got involved with "matter waves." He reasoned that since light had a dual nature (wave and particle) that perhaps matter should have some wave properties also. He stated that the wavelength of matter could be found by:

$$\lambda = \frac{h}{p}$$

where: λ = wavelength of light
p = momentum
h = Planck's constant

He said that for matter, p would be the momentum of a particle of matter.

Huygens (1680) had a theory about light waves but it was not accepted by the scientific community because he couldn't figure out what the wavelength of light had to be. In 1800 Thomas Young was able to perform this calculation and made the concept and theory of Huygens much more acceptable. So, with much work many scientists then got "into the act" bringing forth *wave mechanics* of Erwin Schrödinger, the *probabilities of particle-position existence* of Max Born, and Heisenberg's *uncertainty principle* (1927). It was Niels Bohr who discussed the complementary nature of waves and particles and he said they didn't contradict in theory.

THE CONCEPT OF LIGHT SOURCES AND ILLUMINATION

We know that if we have a light source at some point in our room that the light rays will go out in all directions (assuming no reflectors or focusing) so that each point of a spherical surface in space surrounding the light source will get energy from the source at the rate of:

$$E = \frac{I}{4 \pi r^2}$$

$(\pi = 3.1416)$
(I = intensity of illumination in ergs)
(p = distance to sphere surface)

This is simply the equation of a sphere divided into the available energy in ergs. So the energy is distributed all round the sphere interior as we should expect it to be distributed. It is very interesting to note that this same kind of approach is taken when calculating the energy from a radio or radar antenna that is nondirectional. The equations are the same as for light.

So, some texts tell us that if we have two sources and two spheres we can adjust the size of the spheres or the distance of the source from the sphere surface until our eyes tell us that the energy falling on that surface is the same for each of the two sources. Actually this may not be true as our eyes may not tell us the truth, but visually the illumination is the same. At that time there may be a difference in the size of the spheres or the radius of the spheres and so on. But, when we get this condition, we tend to say that the energy from the source is the same for both of these resulting spheres. They each have the same illumination as the other. It is stated that the energy in the illumination can be expressed as a simple fraction:

$$\text{Energy} = \frac{I}{r^2}$$

r = sphere distances
I = intensity

and this is the inverse square law of radio-radar and microwaves with which we are all familiar. The energy falls off proportional to the square of the distance (r) over which it travels.

Since our eyes may "lie" to us about the intensity of illumination, and this fact was well known by earlier scientists, they decided to do some experiments to remove this difficulty. They devised some experiments (involving the Lummer-Brodhun photometer) that still required visual comparisons but were more accurate. Today we use photoelectronic devices (light meters) which easily and accurately measure the intensity of light from any type of source, even reflected light rays.

Why all the worry about equal intensities of light? Well, if you are going to create or obtain a light source for an optical fiber system, and you have some idea (by other calculations) as to what your losses might be for that frequency of light, then you might work backwards and find out what the intensity of your light source has to be to get that required energy through the fiber-optical system, over a given distance, with a given type of optical fiber. It is nice to be able to use calculators and computers to work out all the values of things you need to have to make such a system work correctly. That way you don't have to do endless experiments which are costly and time consuming to get the answers. You work out the mathematics of the situation and components and problems and then you conduct some experiments which are valid and require the minimum of time, cost, and energy, to prove the validity of your conclusions.

Sometimes, as we have previously indicated, we need light of a given color—which means a given wavelength—to "match" our optical-fiber system. Basically we are told that although light is composed of hundreds of "shades" of color (maybe thousands) there are fundamentally just six basic colors used that we can observe easily and effectively. These six colors are red, orange, yellow, green, blue, and violet.

We have also said that "white light" is multi-colored light, so now we ask ourselves if it is possible to separate some colors from the white light. The answer of course is yes, and we can use a prism to do so. White light into one side of a prism will emerge as several beams due to the frequency diffraction deviation, or color separation, and wavelength refraction due to the different index of refraction for the different frequencies in the material that makes up the glass prism. Some light frequencies (colors) will be refracted more than others, so the basic color spectrum appears. The basic color spectrum is the six colors mentioned, and Fig. 5-8 illustrates a normal type dispersion which is not to any scale but is just illustrative. You do not find the colors so markedly separated in an actual experiment. They tend to blend together. But the sketch does show how the different frequencies will be refracted one more than the next one and that is what we are interested in at the moment. It now seems that if we do not have a source of *monochromatic* (one frequency) light (such as obtained from a laser) then we might use a prism and a white light source to get a given frequency (color) for optical-fiber experiments. It would take some "doing" but it might be done. The word *dispersion* can mean spreading out the colors of light. So, if we say there is *dispersion* in an optical-fiber system and

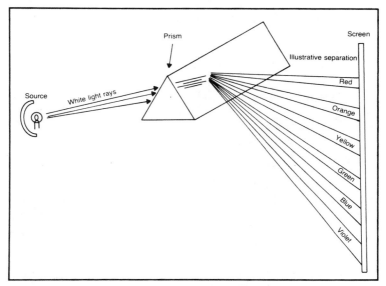

Fig. 5-8. How a prism separates the colors of the spectrum.

this causes losses (we have said that) then what we mean is that the various frequencies are separated in the fiber and some are much delayed or bounced around more than others. We also need to know that there are differences in prism materials. Some may absorb some colors and emit others and so on. Infrared is the invisible part of the spectrum beyond red, and ultraviolet is the invisible part of the spectrum beyond the violet. The wavelengths range from 8000 angstroms to some 4500 angstroms where 8000 is in the infrared region. The angstrom unit is 10^{-9} meter. You might study deeply into prisms and the light effects through them and be astonished at the results.

ONE PRACTICAL ROBOTIC APPLICATION OF OPTICAL FIBERS

More and more we are reading about robots and androids. The "development trend" is to have a computerized machine that will be able to speak and hear and respond to commands or tell you why it cannot respond to them. Also useful is to have machines that can "see." To have a machine "see" is to have it be able to identify shapes and colors and evaluate distances, which is about what we can do with our eyes isn't it?

The Japanese are among the leading developers of Robotic machines (see our book Handbook of Advanced Robotics, TAB book number 1421). The Japanese say they have made a machine that will simulate the five human functions of seeing, hearing, speaking, moving, and manipulating with its single arm as "delicately as a human" can maneuver an arm and fingers. It is said that this machine can respond to a human voice command to go to an object, pick it up, and carry it to a new location. If there are obstacles in its path it will announce (in a feminine voice) that such an obstacle exists and then it probably waits a command to tell it what to do about it ("go through it", "go around it", "move it to one side and resume your main task" and so on).

It is said that there are some 14 to 15 microcomputers connected inside this Robot's body. They are interconnected or interfaced with optical fibers. Of course, we can imagine why this is important. No electrical interference will be picked up by the optical fibers, nor will they be disturbed by noises or electrical or electromagnetic fields around the robot. It is said to "see" very well. Its eyes are made up of some 30,000 optical fiber lines! These are bunched into a bundle and have a moving mirror at the "seeing end" of the cable. Two such eyes, separated in space, can provide distance measurements, identify objects by comparing shapes with memorized shapes in the computer's memory, and identify the location of objects (also from a memorized and preprogrammed software).

Along with the use of optical fibers for "eyes", this robot is said to use an optical character reader, and a kind of radar technology to determine its own position. We suspect that the "radar-type" application may be an extension of the character reading scanning system currently in use in food stores to scan bar-graphs and display or verbally tell us the prices and costs of items. It does not take much imagination to conceive of a system that (given identifying characteristics in its memory) can scan room volumes or other spaces for various "things" and identify them properly.

The use of optical fibers doesn't end here. Planned improvements in a robot consists of placing "eyes" on fingers or fingertips, or arms to give better functioning and more accurate functioning of these members. Again, optical fibers are the answer to transmission and interfacing of elements, objects, and data. As robotics becomes more common the importance of optical fibers in their systems will increase.

Chapter 6
Modern Physics and Laser Technology

One of the prime sources for the frequency of light desired for use in fiberoptics comes from lasers. We assume some general knowledge of lasers at this point and that everyone knows they can be dangerous! **Never look into a laser or LED beam no matter how it is generated or how weak it may seem to be!** Something happens in a very mysterious way in a laser to cause the atoms in certain kinds of substances to emit light rays that are in-phase. This light emission is *coherent*. It may be *temporally coherent* or it may be *spatially coherent*.

Of course, we don't want to imply that a laser light is the only kind of light useful with optical-fiber applications. Our previous chapters help to show us that this is not the situation. LEDs, as well as possibly other sources of light rays might be used. Some are now known. Some, perhaps, are yet to be developed.

WHAT IS THE BASIC PRINCIPLE OF LASING?

It sounds so easy. You just "pump-up" the electron energy in an atom to a higher "state", and then, when it falls back to its "ground state" it gives off light rays in the process. It turns out that in doing this bit of magic, it produces the kind of rays that have all their wave fronts in line. The rays are said to be coherent. They are said to be pure. They are said to be of one frequency only, and not (like white light) composed of innumerable frequencies all jumbled together.

To those who have some radio-radar electronics experience, it is rather like having a single frequency cw note (cw means continuous wave) rather than a jumble of sound produced by something like the old spark-gap type of transmitter. A man named *Charles H. Townes* developed a device called a MASER. This is an acronym for "Microwave Amplification by Stimulated Emission of Radiation." When one considers the increase in range of radar by doubling or even tripling its power output, compared to the increase in range by being able to use a device such as Dr. Townes developed to minimize the internal circuit noise of the front end of a radar system, it is no wonder that this concept had the scientific-engineering world "standing on their ears!" This was a process of taking a very tiny signal, mixing it in a peculiar manner in the MASER device so that it is amplified an untold number of times. The "device" first used was a kind of gaseous ammonia creation or a device using a rubidium crystal. Later other elements were found to be usable such as synthetic sapphire crystals that could be doped with chromium (ruby). Thus the name *ruby laser* came into our vocabulary. What is interesting here is that not only was the type of device used able to amplify signals which were very weak, but it was also able to generate signal outputs in the frequency of light spectrum. It is said that the first laser output was at a frequency of some 6943 angstroms.

We are more interested in the light emitting capabilities of the device we call a laser. So we think back on how the atoms were "pumped-up" so that their energy state was higher than normal and when those atoms returned to their natural (ground) state they would give off this excess energy *multiplied many-fold* in the laser light beams of coherent light. We recall an ellipsoidal chamber (which has, you know from basic geometry, two focal axes) and we position a ruby-rod material along one focal axis and a helix shaped "flash-tube" along the other axis (the inside of the ellipsoidal chamber being very highly polished to provide all the reflections possible). Then if the flash-tube is "fired" using a high voltage power supply triggered by a relay closure so that the flash lasts only a fraction of a second, the light energy in the sealed ellipsoidal chamber is focused by the chamber geometry on the ruby-rod. Reflective mirrors are at one end and a transparent half-silvered mirror at the other end protrudes from the chamber. When this is all properly done, a fractional-second beam of light pulse is emitted from that ruby-rod of such intensity it can instantaneously burn through a steel plate! That ruby-rod was "pumped-up" all right, and there was no doubt that when it was so excited it produced laser light—coherent, monochromatic, and full of energy!

LASER DEFINITION

One accepted definition of a laser is: "A device that produces *optical radiation using a population inversion* to provide Light Amplification by Stimulated Emission of Radiation". Generally, an optical resonant cavity is used to provide *positive* feedback. Laser radiation may be highly coherent either temporally (time-wise) or spatially (position-wise) or both. Also very important here is the concept of *lasing threshold,* which is that level of emission that takes place due to stimulated emission and not to spontaneous emissions.

So we agree, hopefully, that if we can somehow "excite" the atoms in various kinds of chemical substances, we can cause the atoms to change in some manner so that they are not in their normal state or condition. When we remove the excitation and the atoms again return to their "normal" state they will give off some light ray emissions that we call laser light rays. These rays are more power-ful energy-wise, and coherent (in phase and usually single fre-quency) than what we find in "regular" light rays. To understand this action, it seems we still must pursue the concept of "pump-ing-up" the atoms of these selected materials. What do we really mean by these words?

"PUMPING-UP" THE ATOMS

Before we get to the atoms themselves let us consider some familiar concepts related to the absorption of energy. We are all familiar with the thermal switch. This is a metal element that absorbs heat energy and when it gets enough, it changes its physical length, or position, to open or close an electrical circuit (usually). We find these on electric ovens, electrically controlled water heat-ers and so on. The significant concept here is that the heat energy applied seems small, but the reaction of the element to heat is quick and powerful after enough heat has been absorbed. Usually this reaction is a physical movement.

In the laser concept an absorption of energy is also ac-complished but this does not result in anything we can see or hear or feel as far as the chemical element itself is concerned. We see the *result* of this absorption-relaxation, of the atoms in the release of the powerful laser light beams, but we cannot directly observe what happens inside the atoms as this phenomena occurs. We *can* see a thermal switch element begin to change shape, or expand, or whatever, if we watch it close enough and we know its atoms are also being excited by the heat radiations to cause this physical change.

If we examine Physics texts we learn that there is a model of the atom called Bohr's model, which depicts the core of a spherical unit as having a lot of balls called Neutrons and Protons tightly held together (and we don't know with what kind of force). Orbiting around them at given distances (called "shell-distances") are other smaller circular balls called electrons. Three facts are important. The neutrons have *no electrical charge* (neutral = neutron). The protons have a *positive* charge and the electrons have a *negative* charge.

Even with a very elementary knowledge of electricity we know that positive charges are supposed to attract negative charges. We have learned that old maxim "Nature abhors a vacuum" or to state it more generally: "Nature hates an unbalanced condition." So the "nature" of things physical is that the positive and negative charges *should not* be separated. They need to get together to form a neutral pair of no charge—and they do this for us when we have a run-down battery in our car or flashlight! Our batteries are "charged" when the positive and negative elements are separated by chemical or electrochemical (charging) actions. Electrical *current* flows as the electrons try to reach the protons and equalize them. So, why don't the electrons in an atom "fall-down" toward the center of the atom and neutralize the protons there? It is because they are considered to be orbiting—rotating around the core with some velocity, faster as they are closer to the core, and slower as they are further away from the core.

These electrons are then considered like satellites around our earth. We know they are attracted to the earth by gravity. But they don't fall. Why? Because they are moving at such a speed that the distance they fall is the same distance that the earth curves away from the horizontal. See Fig. 6-1 for a view of this elementary concept. Please realize that there can be an orbital path which may be circular about the earth, if the velocity of the satellite is correctly adjusted for that kind of path. An ellipsoidal path is shown, on which the satellite is fastest at *perigee* and slowest at *apogee*.

Back to the atom. The electrons in the Bohr atom are said to be like the satellite. They move around the nucleus at a proper rate of speed so that their centrifugal force equals the attractive force of the core or nucleus and so they continue to orbit and do not fall into the center of the atom as we might expect them to do. We know from electrical basics that the electrons in the outer "shell" are held the

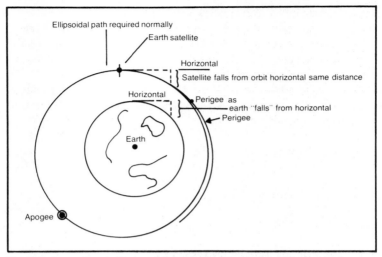

Fig. 6-1. An earth satellite in orbit, is like Bohr's model of the atom.

most weakly in the atom by the core forces, and so under proper conditions they might be caused to move out of one atom to another and then to another, etc. This we know as the "flow of electrical current." When the outer electrons "bond" with other atoms in a permanent manner then the atoms are said to have bonded together to form some kind of chemical-physical substance such as copper, iron, gold, ruby, or whatever.

In the model of the atom as proposed by Niels Bohr in 1913 it was necessary to do some explaining in order to account for some kinds of discrepancies which seemed to exist in the comparison of the electron movement with that of the satellites or solar system planets as we know them. One such discrepancy was that the orbit's radius did not seem to be continuous but seemed to exist in finite "steps" away from the nucleus. We know from basic mathematics that a satellite can be put into an orbit at *any* distance from the earth, given the proper velocity and direction of movement—not so with the electrons. They seemed to have absolutely certain "rings" of orbit into which they could "jump" going out further and further from the core center but they just *could not* orbit between these acceptable "rings" or "shells." Thus we have the concept of *quantum* jumps. It is said that when electrons are "excited" (meaning they had been given some energy in some manner and absorbed that energy) they would "jump" to a higher orbit.

Some problems still had to be resolved. For one thing according to classical electromagnetic theory an orbiting electron would have to constantly lose energy by radiating light (high frequency energy which is related to its orbital period) and so should finally lose enough energy that it would "fall" into the core center or nucleus of the atom. But, this doesn't happen! So we come to the idea of the *"ground state."* That is the orbiting distance at which no more energy can be given off as light rays by the orbiting electron! Also, *Bohr* said that an electron would only lose energy when it "jumped" from a higher level of excitation to a lower level of excitation. Now realize that when we say "lose energy" we mean that this electron will emit rays of light-frequency energy and that idea is important to our concept of the emission of laser light. So, Niels Bohr, genius that he was, was able to explain the stability of the hydrogen atom and its distinctive pattern of sharply defined wavelengths of light which this atom radiates.

HOW ARE ENERGY LEVELS EXPRESSED?

There is a very simple equation:

$$E = \text{Allowed Energy} = \frac{-E_o}{n^2} = \frac{E_o}{N^2}$$

where E_o is a constant number and (n) is a positive number called the principle quantum number, and this then designated the energy levels of the electron. (E_o has a value in electron volts of 13.6.) This is the value an electron gains when it is accelerated through a voltage potential of 13.6 volts. When (n) is equal to (1) then the electron is said to be in the *ground state* as we have just defined it according to Niels Bohr. Thus in the ground state the hydrogen electron (a hydrogen atom has one electron and one proton) which is the basis for most of these kinds of studies, has an (electrical) energy level of -13.6 electron volts. The minus sign shows that the force is opposite the attractive force produced by the nucleus which is positive. Higher energy levels will then be stated by a larger value for the letter (n) and in the limit, as (n) approaches infinity, the energy level approaches zero, which is as we should suspect, because then it is so far away from that core that little or no attraction exists between it and that core.

So, what we are beginning to realize is that if we are going to "pump-up" the energy levels of atoms in a given substance, we must somehow cause the electrons of the atoms to "jump" to higher orbits or into other shells by adding energy to them. The light energy we added to the ruby-rod laser in the ellipsoidal chamber is one method. There are other methods we shall investigate.

We also now should have an intuitive feeling that once we have added this extra energy to those atoms in the given substance, that the electrons, which are not in a normal state of orbit for that substance won't like to stay in this excited state, and so they tend to "jump" back to their normal orbit distances. In doing this, they give off energy (light radiations) and that is exactly what we want them to do. Also, since the frequency of the light depends on the orbiting frequency or rotational period of the electron around the nucleus (and in a given substance all electrons are assumed to have the same orbiting frequency) the light rays have the same frequency (are coherent) and we want that also. We also assume that the rise to an excited level, and the return to the "normal" level or ground state level where no emissions take place, will be accomplished simultaneously by all atoms in the substance. We then assume that what we get from the process is a beam of light (coherent) and with much energy contained therein. We get a laser beam of light.

WHAT IS AN ELECTRON?

Albert Einstein theorized that light could interact with electrons in a manner so that the light itself might be considered "bundles of energy" or quanta. This helped to explain the photoelectric effect as we know it. In this concept the light rays are considered to be particles that do not have any mass (if you can imagine that) called photons and these have a movement which is the speed of light itself. Each of these particles has an energy content that can be expressed as:

$$E_o = \text{Photon energy level} = hf \text{ joules} + K_{max}$$

(h) is Planck's constant = 6.625×10^{-27} erg-sec.

(f) is the frequency of vibration or oscillation and it must be a quantum number that is a digit and not a fraction of a digit: 1,2,3, etc.

K_{max} represents the *maximum* kinetic energy that the photoelectron can have outside the metal surface.

Joule dimensions $= \dfrac{(k_g \text{ m}) (\text{m})}{\text{sec}^2}$ Newton-meter

K_g = kilogram, m = meter.

THE ENERGY OF THE ELECTRON

It is stated in physics texts that the electron has two kinds of energy. The first is that due to its movement and is called *Kinetic* energy. Second, it has energy due to its distance from the core or nucleus and this, like anything raised up above an attractive surface such as a metal ball carried to the top of a skyscraper, means *potential* energy. The longer the fall the larger the potential energy so long as the attractive force is present. Also, it is now a concept that electrons may not be just particles but may be "clouds of energy" and they cannot be well defined in position or orbit either. It is currently popular to say that an electron "cloud" has a *"probability"* of being in a given space at a given moment of time. What we are saying is that we really don't know very much about the inside working of an atom even as of this date. We know a lot about *effects* but since we can never observe an atom's interior, at least with instrumentation currently imagined or fabricated, we really don't know what is there or exactly how it all functions. The reason we can't measure accurately is because the measuring process itself would destroy the accuracy of measurement! That is how delicate this operation would be. It is of interest to note however, that some very able scientists at MIT have succeeded in separating a single atom for "effects" study. We refer to Klepper, Littman, and Zimmerman in their work on *highly excited Atoms* (Scientific American, May 1981). Thus there may be hope that in the future still more exciting developments to astound classical physicists will occur.

A SMALL REVIEW OF PLANETARY- SATELLITE CONCEPTS

Since we are much into satellites these days, and we have goodly knowledge of our own solar system and the planets revolving around the sun, let us review the concepts associated with these realities and perhaps that review will help us with further understanding of the inside of (Bohr) atoms.

Two masses will attract each other with what is called a gravitational force. This force can be accurately measured and is stated to be, in the centimeter-gram-second system of units (g) = $6.673 \times 10^{-8} \frac{\text{dyne-cm}^2}{\text{gm}^2}$ (or about one 15 millionth of a dyne) for the attraction between two masses, each of one gram weight, that are one centimeter apart. The equation for the attraction of a satellite to the earth is:

$$\text{Attractive Force (gravitational)} = \frac{\text{(g) Mm}}{\text{R}^2}$$

We see that the force varies *inversely* with the distance between the two objects, the earth and the satellite mass.

There is another equation relating to the velocity of the object in orbit. It is:

$$\text{Velocity required} = (V^2) = R_o\ (g)$$

Where (V) is meters/seconds and R_o is earth radius in meters and (g) is the gravitational constant mentioned earlier but in the appropriate units of meters-kilograms and seconds. Of course the British units can be used so long as the units are consistent. The velocity stated here is the speed required to maintain an object in a *circular* orbit at the earth's surface.

The equation can be stated in another way:

$$\frac{mv^2}{2} = \frac{(g)\ Mm}{2\ R_1}$$

Where R_1 is the distance from the earth's center to the satellite orbiting above the earth. This equation is quite interesting as it shows that the *kinetic energy* (mv^2) needed to maintain the satellite in orbit *decreases* as the satellite gets farther from the earth. Is that not what we have said about the electrons in an atom? The further out they are from the core, the weaker they are held or bound to the core? However, the *potential energy* (PE) of the satellite is related to its distance from the earth by the expression:

$$PE = -mg\frac{R_o}{R_1}$$

R_o = earth radius to center
R_1 = satellite radius to center
m = satellite mass
g = gravitational constant

At an infinite distance from the earth the PE is zero since the satellite is so far away it is no longer attracted to the earth. So it is with the electron mass-cloud. When it is far enough away it also is *not* attracted to a given nucleus of a given atom.

WHAT HAPPENS IN THE ATOM?

We are now interested in a kind of matter-antimatter concept. We assume that the electron is a cloud of energy and that a *photon* comes along and collides with it. By definition the photon is a positive charge and the electron cloud is a negative charge. so what happens? Do we have a matter-antimatter situation? If we did there should be an explosion, a radiation, a *something*, and then nothing left. Here is where we must be very careful. What seems to happen, as evidenced by tests and such, is that *the electron absorbs the photon,* and in doing so increases its own energy level. It gobbles up the photon like Pac-Man gobbles up the ghosties. *If its own energy level is increased enough,* it moves out, farther away from its nucleus to another shell or ring or orbit and it is considered to be in an excited state. What that really seems to mean is that its energy level is increased—but only for a short period of time. It cannot retain that new energy level and so it must give out with something (energy of a different kind perhaps) so it can return to its *stable and normal* energy level condition. What it gives up is a light ray which also might be considered to be a photon under some circumstances! Tests show that the electron-cloud might give up just one photon as it returns to its original orbit or it might give up several photons through a series of intermediate permissible orbits. Note that it cannot just orbit anywhere. It can only orbit in those integer unit levels we mentioned earlier.

It is convenient to think of the electrons in an atom as being on the steps of a staircase as illustrated in Fig. 6-2. Notice that the "steps" are not equally spaced. We have used more than the usual number of steps (3) for clarity. Consider that an electron may fall from the highest step to the one below it (from step 3 to step 2) and emit a burst of energy and then "fall" to the "ground" state. This means that when "pumped-up" the electrons are given the energy sufficient to move them from an orbit that is closed to the nucleus to an orbit they usually would not have. They give up that added energy and drop back to the neutral (ground) state orbit.

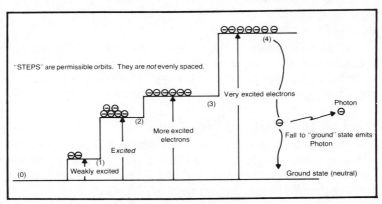

Fig. 6-2. An illustrative case of "excited" electrons.

Many times we read of an "inverted" population in connection with the number of electrons in a given orbit. What this means is that the number of electrons present at that time are more than would normally exist in that shell at that distance from that nucleus. An "inversion" usually means to turn something up side down. Here, it means to have an over abundance, or too many, or more than usual, yet it is called a *population inversion!* Well, so be it.

Another point worth clarifying is the name MASER (Microwave Amplification by Stimulated Emission of Radiation). This means that a *microwave frequency* will be generated or amplified by this process. The word *laser* changes the frequency from microwave to that of light. Thus we have Light Amplification by Stimulated Emission of Radiation. The only difference in the two words is in the frequency of output being generated.

Atoms might be considered to be a very complex "world" and the electrons might be thought of as clouds of energy (especially in the quantum concept). Now we can imagine that we have lots of atoms that are turning and tumbling and twisting with respect to each other as inside each of the electron clouds they "vibrate." It is said that when an atom is "excited" by gaining energy either by collision with another atom or by absorbing photons of light (and many astronomers think now that the absorption is of infrared light frequencies) the energy of that atom so excited must increase by a "quantum jump" to the next or many steps higher in the energy staircase shown in Fig. 6-2. Notice that since the stair steps are steeply fronted, the electrons or excited atoms cannot exist anywhere except on the steps or at discrete levels above the ground state. The change in energy level cannot be gradual but must occur in "jumps" (quantum jumps is correct terminology).

When an atom or molecule "drops back" to a de-excited state or a "less-excited" state, its energy difference must be dissipated either as a radiation or as a photon, or in a collision with another atom or molecule which then absorbs that energy. Nature doesn't like to have unbalances, and so it doesn't like inverted populations, nor does it like over-excited atoms.

Now we must make an assumption. Suppose that some energy of the photon type, rather weak but strong enough to change the energy level of a substance, is caused to interact with that substance. What happens? The substance becomes "unbalanced" and "excited." Now, remove the "pumping" energy and let the atoms of the substance seek their own balanced levels of energy. What happens? Suppose that in seeking this return to the balanced level, the atoms do so with a release of more energy than caused them to become excited! If we then assume that "photons-in" means "photons-out", you know by intuition that we are going to get some amplification of that energy we put into the system. If the input is light, we get out a stronger intensity light. If the input is a radiation, we get out a stronger intensity radiation.

You cannot get something for nothing. The energy into the system must create the excited state right, but we cannot get more out of the system than we put into it. If this be true, then we consider that we have concentrated the energy input into a focused output, and so over a smaller spread of some dimension it does seem more per unit dimension than we put into the system. There may be other ways to rationalize this idea. What we try to consider is the word amplification as associated with the rise to an excited level and drop to a ground-state level of the electrons in a maser or laser-type device! Quantum mechanics tells us that atoms can only be raised to various discrete steps by energy absorption, and then they emit just exactly the right amount of energy which corresponds to the difference in the energy level to which they "drop."

So it may well be (as most believe) that it is the fact that many atoms are all caused to go to the excited state and then drop back simultaneously to a lower energy level that produces this "amplification" effect. An interesting concept associated with this idea is that the atoms might be "pumped-down" to the ground states. That would mean that they have an inclination to remain in the excited state for a while ("while" meaning some time span in some fraction of a second) so that in this "pumping-down" action—and we are not sure what this means although it is mentioned occasionally in the literature—many atoms might be caused to move simultaneously and so we get an increased output, which we call amplification. It may be a somewhat new idea to you, this "pumping-down" of atoms, but now you know about it.

So we are left with the idea that it is possible to excite atoms and then get an increased output from them when they return to the ground state. The next idea is associated with *how* we excite them. We know there are lasers which are large and gas filled, there are lasers which have solid elements in them such as the synthetic rubies, and there are integrated circuit lasers, etc. It is this latter group that now most interests us because of the smaller size and possible use in optical-fiber applications.

EXCITING THE SOLID-STATE LASER

It has been found that lasing action in elements can be caused by the absorption of photons and that means a light intensity that causes the excitation of the atoms. We learn also, that there is a method of simply passing an electric current through some solid-state, transistor-like, elements which can also cause atoms to become excited. In this latter unit, if we provide the current in a proper manner and provide a proper geometry to the transistor-like elements, then we can have "windows" through which lasing beams are emitted. We indicated this earlier in our diagrams of solid-state laser light-emitting devices for use with optical fibers.

We note with interest that energy can be imparted to an atom through the process of *collision*. If we have electrons bumping into one another, energy is imparted and lost in the process. Thus we have a basis for *gas-lasers* using the impacts of two gasses. There are certain conditions that must be met before an energy transference is possible, that is the electrons must exist in certain energy levels around the nucleus so that a collision will cause some to go to a higher state level and some losing energy to go to a lower energy state level, and neither levels must be the ground state. The use of radio-frequency energy can be used to excite these gaseous atoms through a rf discharge tube. This produces an ionization and thus frees electrons to move around and cause collisions. Also, as you'd guess, the use of electric currents might pass through a gas and cause free electrons to "jump around" and so cause collisions. Once a lasing action is started, it is possible to use the laser beam itself, reflecting it back and forth in the substance or gas, to produce a build-up of the laser beam. This *positive feedback* arrangement is used successfully and is mandatory with some types of laser structures.

THE GALLIUM-ARSENIDE LASER

While there are many types of commercial lasers on the market and much information may be obtained about them, we will now turn our attention to the gallium-arsenide laser which is a solid-state type and can be obtained for home experiments. It is a type that seems adaptable to optical-fiber applications.

The gallium-arsenide laser is a good example of a solid-state type of diode which can emit light in the infrared region when an electric current is passed through its junction. It is also called an *injection laser*. The reason for this is that when the electric current passes through the junction between the (P) and (N) materials light is emitted. This light (in the infrared region) is not visible to the human eye. It is of interest to know that the infrared spectrum is said to consist of three regions: The (NIR) region or near infrared region, the (IIR) or intermediate infrared region and the (FIR) or far infrared region of the spectrum. Near is assumed to be that *near* visible light frequencies and *far* is assumed to be that which is at the opposite end of the spectrum of infrared emissions. *Injection lasers* are adaptable for communications applications because they are small, light, and compact, and require small amounts of electrical energy. Their outputs are sharply focused, and they can be "tuned" to certain "windows" in the atmosphere so that absorption is relatively small. They can be switched on and off at the fast rates required for digital communications.

When we consider the "windows" in the atmosphere that will pass the infrared emissions we are considering 0.95 to about 1.00 microns, 1.2 to 1.3 microns, 1.5 to 1.8 microns, 2.1 to 2.4 microns and 3.0 to 5.0 microns. These are, of course, not exact but are close enough to give us some ideas of what the "window" frequencies are considered to be. There are other frequencies for other types of laser light which are windows for sea-water and so on. Weather can have an important part in determining which window *is* a window at what time.

We recall some experiments with a "heat-seeking" head for a small guided missile. It was of considerable pleasure to hold a lighted cigarette some 30 feet away from that "head" or nose cone and have the control surfaces deflect as the head tracked our cigarette. These type missiles are sometimes used against jet aircraft and so their detectors are necessarily required to have windows which "see" the jet exhausts. Of course there had to be developed countermeasures in the form of flares which also were so designed to radiate the "window" frequencies strongly and when these were ejected from the target aircraft, they could possibly present complete confusion to a heat-seeking missile.

Some applications require domes over the missile sensing units and these could be of some material that is transparent to the "window" frequencies and opaque to other frequencies. Glass and quartz might be used for some frequencies and synthetic sapphires, silver chloride, and sodium chloride might be used for other frequencies.

In a previous chapter we illustrated a kind of strip-edge laser light source. In Fig. 6-3 we amplify that concept. This type of laser

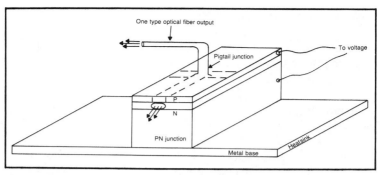

Fig. 6-3. One type of a strip-laser light source.

diode may be constructed from AlGaAs and it is said that its operation depends upon internal reflections along its strip to cause an increase in photon population to a level where emission takes place. One must be very careful with this type of laser to prevent temperature increases that might destroy the unit when it is being operated. Normally there is a feedback circuit used which monitors the output and controls the exciting voltage to prevent undesired temperature increases.

The light emerging from the active area of the injection laser diode is coherent and is in a narrow beam whose size is governed by the active area of the diode. If you study the spectral output of such a beam you find that there is coherent light. This means that when the electrons are "pumped" to a higher energy level and fall to the lower levels emitting photons, they may all do this in exact time phase. Most rise and fall simultaneously, which is desired, and these produce the coherent light required as an output from a laser.

It is possible to visit a radio-parts store and obtain LEDs and request some lasing diodes and with these you'll get circuits that show how to connect them. **Be careful! Never look at the output of any device that might produce a laser light. It could blind you!**

SOME CONCEPTS OF THE SOLID-STATE LASER

It is necessary to couple an optical fiber as closely as possible to the light-emitting part of a laser diode in order to reduce light losses. Any light that does not couple into the fiber is considered to be "lost" or wasted energy. As far back as 1971 some scientists (Burrus of Bell Labs to mention one) developed coupling systems to get the light from a diode device (a LED or a solid-state laser) into an optical fiber conductor by making a solid-state chip on a heatsink and planning its geometry so that an optical fiber "couple" could be run right down into it to the light producing area. The head, associated with the conduction of electric current through the diode, is carried away by the metal plate or "heatsink."

More popularly accepted and currently of great importance was the development of the strip edge emitter of light. In this configuration (shown in Fig. 6-3) the metal contact to the element of the diode is so arranged that it presents a strip or metallic ridge which runs through the diode at the junction of the PN materials. A section is free so that light can be emitted from the junction area. Due to the metallic connection, a very active region of electron-hole dynamics will occur in this type of lasing diode. Connection of this type of light source to an optical fiber is made using a "pig-tail" part of the fiber.

In the strip diode-laser an initially excited photon goes down the strip of metal, and by collision and interaction with other atomic elements excites other photons. Since the ends of the strip are optically flat then reflections take place and so it is considered that the strip acts like a resonant cavity. If the oscillations are properly stimulated according to energy level and frequency, then lasing action takes place and coherent light is emitted. It is of interest to realize that the two strips used in the PN junction are carefully machined so that they are perfectly flat, perfectly parallel, and reflective, and thus form "mirrors" so that the photons will be reflected between them after initially being excited by some free electron (provided by the forward bias arrangement of external current to the diode-laser).

One problem with a lasing diode of this type has been the small amount of power available from it. Now comes the news that Xerox's Palo Alto research laboratories have developed a "ten strip" or *stripe* semiconductor laser which can emit up to 410 milliwatts of optical power. This is about seven times more than possible from the single strip (stripe) laser mentioned earlier.

The new laser is a gallium-arsenide-aluminum diode and it derives its fantastic capability from a new design having ten etched stripes (strips) instead of just one. These are all optically connected (by mirror and reflection and parallelism) so that the resultant output beam is coherent. The beam output occurs at about 825 nanometers in the infrared band. It is expected that this type laser, if improved, and if frequency of the light output can be controlled properly, may replace the helium-neon lasers now being used in document scanners, optical data stores and systems, and on printers and so on. It is of interest to know that most helium-neon lasers have an output of only about 50 milliwatts at this writing.

We have mentioned laboratory experiments that are so refined that single atoms can be separated from molecules in order to study them. We note here that this concept has become an important tool in the study of atomic energy, especially the separation of U-235 from U-238. Individual atoms of these two isotopes are said to differ from each other by less than a septillionth of an ounce, or the weight of three neutrons.

Currently, in order to separate the two atoms an involved process has been required. Refined uranium is mixed with flourine, converted to a gas, compressed and pumped into one side of a chamber divided by a sieve-like membrane. The U-235 atoms zip through the holes in the membrane with some ease while the U-238 atoms come through but not in the same numbers. But, after going through this process a good many times the concentration of U-235 is pretty good and is above that of U-238 and in nearly a usable form. Another way of separating the atoms is to take advantage of the fact that the U-238 atoms are heavier than the U-235 atoms and so if one puts the gas into a centrifuge and whirls it fast, the U-238 atoms tend to move out away from the U-235 atoms and so the U-235 atoms can be recovered from the center of the machine area.

We now find that lasing action can be used to separate such atoms. If a beam of light can be focused on vaporized uranium this will cause the emission of electrons and make the U-235 atoms ionize (become positively charged). In this charged condition, these atoms can then be influenced by an electric field (like charges repel, unlike charges attract) and so the U-235 atoms can be attracted and collected in a very pure state.

There is a "catch" to everything. The "catch" here is to have the correct laser light frequency. You just don't stimulate uranium atoms with just any frequency of laser light. The scientists won't indicate the frequency of the lasing light, but they do say it is a kind of "reddish" color. It was of importance to note that in the machine that does the separation, a green light is used to stimulate atoms to produce this "reddish" light, which causes the ionization of the uranium atoms.

EXCITATION OF ATOMS

As we have mentioned, atoms can be excited to various states by various means (light, rf energy, electrical currents, heat changes, and so on). Such atoms when excited have electrons in the outer shells that accumulate in such a manner that the atom is, essentially, unbalanced. Such atoms then tend to return to the *normal* state and when doing so, cause photons of light or, perhaps, emissions of some other type, to emit from the atoms. When all emissions take place in a proper timing frame then we have what is called *coherent light*. We have indicated that atoms become more excited as more energy is absorbed and electrons move farther away from the nucleus of the atom. When in the most highly excited condition, electrons may actually break away from the bonding of the atoms and drift in some manner from atom to atom. The atom they leave is then said to be ionized, or have a net positive charge. Ionized atoms are often used to excite other atoms, being injected into, say, a chamber filled with a certain gas or gassified material by means of an *injector* rod. Any person who has had experience with radio-frequency wave generation and transmitters or television high-voltage power supplies probably has had the experience of seeing the blue, ionized emission, which can occur from any sharp pointed object connected to the generating source. You have also probably smelled the odor, kind of a burnt smell, which is released simultaneously with the *arc* or *arcing*. In days past you probably, if you are electronically oriented and have experimented with such units, have used some kind of spray which is insulating to a very high degree, to "smear" the point, after, perhaps, trying to physically smooth it down somewhat, to prevent such arcing? In any event that blue emission was the emission of ions, positively charged atoms, which had lost their balancing electrons.

Thus we come to the idea of highly excited atoms or Rydberg atoms as they are called. These atoms are special being very large in size (for an atom), they have a very long life, they are very much influenced by a magnetic field, even if it is a weak field. When highly excited, they can have an electron orbit outside the fields of other electrons, but still be held to the atom by the attraction of the core and possibly the remaining electrons. This atom also can be explained by the Bohr Quantum Theory: i.e., atoms exist only in certain energy-level states. The electron cannot, ever, spiral into the nucleus, it can only lose energy by "jumping" from a higher energy level orbit to a different, lower energy-level orbit. It will give off excess energy as an electromagnetic radiation in the energy-losing process. It has been found that the allowed energy level states can be described mathematically by:

$$\text{Energy} = E = \frac{-13.6 \text{ electron volts}}{\text{principal quantum number}}$$

and, again, one electron volt is the energy gained when an electron passes through, is accelerated through, one volt of potential. When the principal quantum number is (1) the level is the "ground state" or has a potential of −13.6 volts. The minus means this force is opposite the nucleus. The quantum number (n) must be an integer. It has been found that the radius of a Bohr orbit of an electron is proportional to (n^2), and for the Rydberg orbit (n) may be between 10 to 100 according to MIT scientists. This may be of importance to us in that, as Bohr surmised, the energy emissions from an electron which "jump" from a value of (n) near 100 to the next lower level (from 90 to 89) will be much smaller than those emissions which take place when the electron "jump" from a level of n = 2 to n = 1 or the ground state. There is a smooth change in emission levels when (n) is large and this is said to contrast with the energy emissions from atoms where the "jumps" are made with small values of (n). Perhaps this could be important in lasers which operate continuously versus those that are pulsed. It must be noted that in the Rydberg state the atom is very close to being in the ionized state as that "far-out" orbiting electron might be torn away from the Coulomb attraction of the core at any time.

THE ALKALI METALS

These are said to be the work-horses of atomic physics. They are: lithium, sodium, potassium, rubidium, and cesium. They are used because they are easily converted into a gaseous state and because their spectral absorption lines are at wavelengths easily generated by laser light. The alkali atoms may be excited with pulsed, tunable lasers that can generate highly intense flashes of light for just a split second. The light, of course, is monochromatic (one frequency). Using three lasers in one experiment conducted by MIT scientists, two pulses of light got an atom to an intermediately excited state, and another pulse from a third laser boosted the excitation such that the Rydberg state of high-excitation was accomplished.

WHEN CAN AN ATOM ABSORB LIGHT ENERGY?

According to Neils Bohr an atom *can* absorb light energy *if* the frequency of the light, when multiplied by Planck's constant (h) (which has a value of 6.625×10^{-34} joule-seconds) is equal to the energy *difference* between the initial state of an electron and the excited state of that electron. It is interesting that in some experiments when light from a lamp is passed through a gas and then spread out with a spectroscpe so the individual (colors) frequencies can be seen, there occurs dark lines at those frequencies where the light wavelengths have been absorbed. This will occur if the energy at those frequencies satisfies the Bohr relationship.

In the concept of a maser we have to raise the energy level of the atoms or molecules to a particular quasistable state and then stimulate them to fall back to a lower level. The energy they give off in doing this is an amplification of the input energy. We can suppose that the word "pumping" comes from this kind of action: the raising and lowering of the energy level of atoms so they will give off energy, and we could consider the *pumping* concept to be such that it adds energy—like a pump produces output pressure, and then draws energy like a pump "sucks-in" some kind of intake. It takes on the dimensions of a cyclic kind of process and electromagnetic radiations in an oscillatory state would seem to provide the right kind of energy that is required to do the "pumping" job.

Again thinking of the "amplification" process, we note that a molecule at energy level 2 will give up that energy and it takes only a very weak "down stimulus" to cause it to return to the energy 1 state. It gives off a quanta of energy in doing so. The quanta it gives off then triggers other atoms in the same condition to emit energy and so ultimately the energy output is millions of times larger than that which was required to get that first molecule (or atom) to return to state 1 from state 2. It is said that an atom is "inverted" when its energy state is like that just described. It wants to be at energy state 1, but in some manner (radiation or collision or whatever) it was sent to level 2 or higher and it finally wound up at level 2 needing just a tiny bit of down energy to get back to level 1. Then it is said that it no longer is in an inverted state.

It would not be proper to discuss the energy states of the Bohr atom without considering an energy level diagram of the basic hydrogen atom. This atom, as you know, is the basic one because it has one electron and one proton. We also know that every Bohr energy state for the atom is designated by an integer (non fractional) called the principal quantum number of that energy state. An energy level integer of 2 is indicative of more energy than an energy state of 1, which we know now is the ground or neutral, or normal, energy level associated with the orbit of the moving electron cloud. We now realize that when an atom is changed in state the difference in energy is considered to be in the form of a radiation, either absorbed

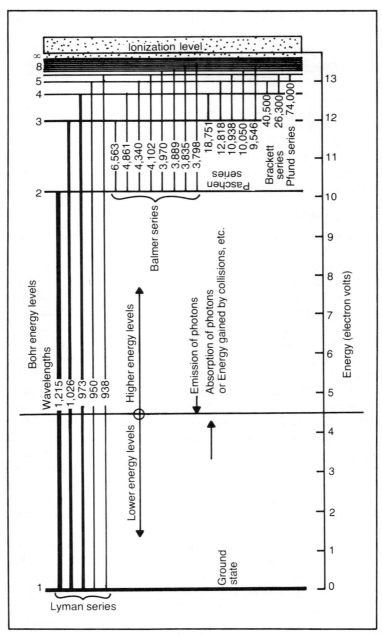

Fig. 6-4. A basic energy-level diagram (Bohr).

205

to increase the energy level of the atom, or emitted in order to permit a return of the atom to its normal state, or toward its normal state. The energy of the quantum emitted is directly proportional to the frequency and inversely proportional to the wavelength. The wavelengths of the atom's radiation are stated in angstroms which is 10^{-8} centimeter length. Figure 6-4 illustrates a basic energy-type diagram of atoms, and the wavelengths of the atoms expressed in angstroms. You will note the classical names of the spectra series such as *Lyman, Balmer, Paschen, Brackett* and *Pfund*. The Lyman series is associated with the lower frequencies or longer wavelengths, while the Pfund series is associated with the highest frequencies or shortest wavelengths.

THE DYE LASER

This type of laser emits light that exists only within a narrow range of wavelengths, but it can be tuned to operate over a relatively broad band. The lasing medium is a fluorescent dye that emits light rays or photons over a rather continuous spectrum when this material is "pumped" or energized by another type laser. Usually an optical cavity is used to confine the emissions to a rather narrow band or even to a particular frequency. The pumping is done by a laser that emits an ultraviolet light and this, in turn, might be guided into the cell containing the dye by means of a quartz lens. It is of interest to note that when considering the forces in the atoms as electromagnetic radiations one can use the concepts of *standing waves* and single frequency generation and absorption by means of properly designed cavity resonators and still utilize some of the principles of optics such as beam splitters, birefringent crystal actions, mirrors, polarizing filters, diffraction gratings and such to conduct experiments and make determinations and create useful devices.

In the dye type lasers the molecules are certain kinds of dyes that normally have a tendency to absorb light photons easily and thus raise themselves to highly excited states. When a given wavelength is repeatedly passed through the dye laser being "pumped" by another laser of some kind, say a nitrogen laser (which is emitting ultraviolet radiations) the dye laser beam gains in strength and is "amplified" tremendously. Normally, to do this a diffraction grating is used to separate out the desired frequency and then a reflecting glass plate at the other end of the device causes the beam to pass back and forth through the dye many times.

CONSIDERING SOME LASER TERMS AND DEFINITIONS

As in all of science one of the keys to understanding is to be able to determine what scientists and engineers are talking about. In mathematics, for example, if you don't know what a "transform" is, you are in trouble. So it is in the field of laser sciences. In modern terminology here are some aids to learning.

Fabry-Perot Interferometer. Used in dye lasers to separate given wavelengths, it consists of two partially reflecting surfaces with a precise spacing through which radiations pass.

Etalon. A specific design of the Fabry-Perot interferometer made from a single glass plate with its two parallel surfaces coated with some partially reflecting materials. Light rays of different frequencies passing through this device take different paths and so some rays combine to reinforce themselves while others tend to destroy themselves by out-of-phase mixing. The phase is a function of the paths taken.

Spectroscopy. A means of causing the different wavelengths of light to separate due to traversal of some kind of medium and so produce a "spectrum" of light and dark (reinforcing and destructive) combinations of the frequencies. Most good physics texts discuss this concept in detail. There are many ways to accomplish this phenomena (Doppler spectroscopy, polarization spectroscopy, photon spectroscopy and so on). A diagram to show the different wavelengths as related to atomic energy is shown in Fig. 6-5.

Gluons. An assumed carrier of a basic attractive force in an atom. It is called the strong force or the color force. It is useful in atomic theories and hypotheses and considered to be a massless type of "particle."

Quarks. The assumed basic elements of protons, neutrons, nutrinos, and mesons etc. A quark is said to have the property of *collector charge* and any such particle can then be said to be able to emit or attract a gluon. Quarks are theorized to be held together by gluons.

Glueball. The resultant of two gluons occurring when an

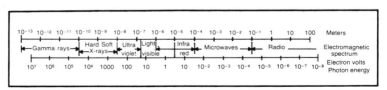

Fig. 6-5. A concept showing wavelengths and photon energy of high-frequency radiations.

exchange with other gluons occurs. A theoretical concept and not experimentally confirmed as of this writing.

Neutron. A mass without charge found in the nucleus of the atom along with the protons that make up the positive charge. Normally, the charge between the protons and the electrons is balanced so the net charge of the atom is zero. If an electron is lost, the atoms becomes *ionized* or positively charged. If a proton is lost, the atom becomes negatively charged, assuming the electrons originally there remain in orbit (Bohr's concept) about the nucleus. Neutrons cannot be influenced by electrostatic or electromagnetic fields or magnetic or electric fields. They are said to possibly be influenced by a particle called a pi meson.

Pi Meson. A particle discovered in 1947 that is said to have a mass about 275 times that of an electron. It is said to possibly interact with both the neutron and the proton at the core of the atom.

Photon. According to some theorists (Yukawa) a constituent of the atom which has a rest mass of zero, as does the gluon. A photon is a particle that carries light energy. It is a part of the concept of light. *Einstein assumed* that the energy in a light beam travels through space in concentrated bunches called *photons* whose energy is given by:

$$E = h\nu$$

where (E) is the energy in electron volts, (h) is 6.57×10^{-34} joule-seconds and ν (nu) is frequency in cycles/sec. of cut-off, or stopping, of emitted electrons in a photoelectric effect. This is the cut-off frequency which is different for each emitting surface. The quantity $h\nu$ is the energy of a photon. A photon cannot emit or absorb another photon.

Color Force. A concept associated theoretically with the binding force of the atomic nucleus. It is currently theorized that the photon is possibly the carrier of an electromagnetic force that holds the atom together. But, in some current theories the *gluon* is thought to be the carrier of a so-called *strong force* or *color force*. It is then said that a gluon (associated with the atom-of-light analogue) has a *color* charge. (See Kenzo Ishikawa, Scientific American, Nov. 1982.)

Hadrons. A particle proposed by Murray Gell-Mann and George Sweig of CIT. which is subject to the "strong force." In their theories the fantasy of colors associated with forces predominates and "flavors" of colors are not unusual terms.

Up-Flavored Quark. An up-flavored quark is a part of a *hadron* in this theory, and is designated by the symbol (u).

Down-Flavored Quark. Also a part of a *hadron* and is designated by the symbol (d).

Strange-Flavored Quark. A particle which is theorized to explain certain mathematical long-lifetimes of some types of particles within an atomic force structure.

Anti-Quarks. Particles postulated to have opposite characteristics to the normal quark concept.

Quark "Flavors". Names associated with two more concepts: the *charm* (c) and *bottom* (b).

From the above listing we begin to get some idea of the strange manner in which theoretical physicists use (perhaps with tongue-in-cheek) various names of other well known things or phenomena or physically sensed qualities to, perhaps, assist in the explanation of the new modern theories of what makes up atoms. They constantly search for the real basic explanations to "What holds an atom together?" and "What are the constituents of the atom beyond those currently measureable or observable?"

To assist these "magicians" in their fabrication of (mathematical) models and other devices to assist in the explanation of theories, which may or may not be of an advanced nature, we find that the *"color"* concept is helpful in that one might diagram such objects as, say, *glueballs* in which we find *gluons*, and as theorized, we know that each of these has a force called *color*. These *colors* are a way to designate that each of these "things" have some unique mathematical properties. They do *not* mean the usual idea of color as we associate this word with blue or red or whatever. So, it is nice and "handy" for the mathematical physicists to be able to say "That gluon emits a *blue* gluon and so it becomes a *red* gluon." They assume that one gluon can emit many gluons and that the combinations of the total number in an atom make it colorless, or as one might think, it has some constant energy state although there is interaction *inside* this "world." The number of gluons in a glueball has not be suggested or well defined in physics.

Fermion. Particles which have a spin (angular momentum) of ½ integer are called *fermions*. These are electrons, protons, neutrons, and quarks. It has been stated that all electrons have a characteristic *"spin"*, i.e., an angular momentum about some axis, and this has a value of:

Electron Spin Momentum $= 0.52723 \times 10^{-34}$ joule-seconds

Whether the particle actually spins, or whether this is just a loop of electric current a *spinning* electric current-loop) may not be precisely known. What is known is that such spins give rise to magnetic moments or angular momenta. The *alpha* particles and the *pions* are said to have zero spin angular momentum.

It is considered, currently, that in atomic and nuclear physics the "elementary particles such as electrons, protons, and neutrons have an angular momentum associated with an intrinsic spinning motion, as well as motion about some external point. It is also recognized that the spin values can take on only definite discrete values rather than a whole series of smooth transitional values. Thus angular momentum as considered here is said to be *quantized*.

If we employ the "right hand rule" of vector notation we can present an elementary concept of a small particle with "spin" and angular momentum. See Fig. 6-6. The classical law of the *conservation of angular momentum* states that the sum of the external torques must be zero. This may or may not be considered by advanced theories in particle dynamics of the atoms.

THE FUNDAMENTAL THEORIES

The fundamental theory of electrodynamic interaction of charged particles in an atom is called *quantum electrodynamics*. In this concept the idea is expressed that *the force* between two electrically charged particles can be accounted for by an exchange of

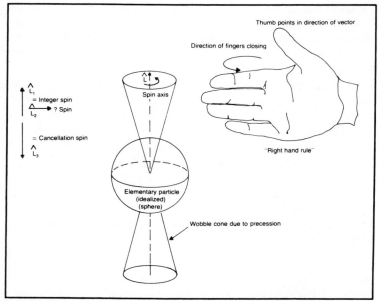

Fig. 6-6. Idealized concept of a particle with spin and angular momentum.

photons (the photon itself being electrically neutral). This theory does not account for the change in charge which may exist from a particle emitting a photon.

The basic theory of the concept of a *color force* may help. Again we mention that *colors* as referred to in this concept and in the modern atomic theory considered by some advanced scientists, is not a color at all as we think of it. *Color* really means a mathematical statement of some kind, and complexity, which, in turn, has some significance with regard to size, mass, charge, type and so on of the particles. By "manipulating" the mathematics they say they can illustrate changes of such objects as gluons, from a "red" color to a "blue" color.

In a way this is a nice idea to describe what might be absolutely confusing (advanced mathematics) in terms that are only a little confusing! We refer to laser light as "red" or "green" or "blue" or "violet" because we can perceive it as having this frequency of oscillation, which our eyes, nerves, and brain then interpret as a "color." One must be careful, then, to know *when* we are associating color statements with *what* kind of theory, or particles, or objects, or light beams, or what? We must also have some idea of what the word "color" means *in that language frame*. In a way it is like the American words "Bare, Bear," (and others) which sound alike, may be spelled alike, but whose meaning depends upon how they are used and in what context.

What we are finding in these pages, we hope, is that the atom is not as simple as one is often lead to believe. Also that in atomic physics as the various particles and forces are discovered and better understood, theories of far reaching importance may be forthcoming. For example, we know that laser beams are rapidly attenuated by atmospheric molecules. Research may find a way to overcome this problem!

MORE DEFINITIONS AND CONCEPTS OF MODERN ATOMIC PHYSICS

We have listed a definition of a *quark*. Now we expand this idea and state that an "up" quark has an abbreviation of (u). The "down" quark may be designated as (d). But these states are called *"flavor"* and that sounds like a tongue-in-cheek statement by some very advanced people. In this concept the proton is said to be made up of (uud) and a neutron is considered to be made up of (udd). It has been postulated that a third kind of *quark*, which has strange properties (long lifetimes compared to others) be designated (s). With these kinds of designations the scientists become very happy for they can describe various kinds of elements or particles in this kind of symbology and it becomes a kind of "short-hand" way to write things and then manipulate the written abbreviations in a mathematical methodology.

The idea of color was first proposed to make the Quark model "jibe" with the *Pauli Exclusion Principle*. This principle says that any two *fermions* cannot share the same quantum mechanical state, which means they cannot have the same energy value, nor spin value, or any other numbers which are used to identify one fermion.

The fundamental theory of the electromagnetic interactions of particles is called *quantum electrodynamics* (abbreviated QED). It used to mean; "The statement, hypothesis, or whatever is now considered to be proven. If you are a modern physicist it means quantum electrodynamics. Learn the new terms and abbreviations and definitions of science—some are very peculiar!

QUANTUM CHROMODYNAMICS

This is the basic theory of that color force we've mentioned. It, of course has an abbreviation: QCD. The mathematical framework for this theory was developed in 1954 by C.N. Yang of the State University of New York, and Robert Mills of Ohio State University. QCD states that one "colored" particle interacts with another by the exchange of gluons. It is said that because there are three colors that predominate, the theory of QCD is much more complicated than that of QED. QED is the theory most of us are familiar with. The gluon colors are said to be blue, red, and green. Sound familiar to the television specialists? But remember, always, these aren't colors at all as we know them. They are *mathematical* expressions or statements or practices or procedures. Among the elementary particles which have been found in an atom are Leptons, six of which are known. One of these is called the electron, another is called a muon and a third is called a tau. These are like electrons except for mass. The muon is said to be 200 times as massive as an electron, and the tau is said to be 3,500 times more massive than the electron.

ON THE ANGULAR MOMENTUM OF LIGHT

We have mentioned the angular momentum of various constituents of an atom. We are interested in light, primarily, in this text and so let us now consider the idea of momentum as associated with light rays or photons. Light rays can deliver linear momentum to an absorbing screen or to a mirror in accord with the classical theory of electromagnetism. We also know that light rays can be polarized. Thus when we consider *circular* polarization it is not difficult to imagine that angular momentum might somehow be associated with this circular polarization. The circular polarization being considered by us to be some kind of rotation of the $\hat{E}$ vector of the electromagnetic representation. That this idea was possibly true, was proven by Beth in 1936 who showed in his experiments that when circularly polarized light is produced in a doubly refracting slab, that slab will experience a reaction torque.

So it was found and hypothesized that if light photons carry away some angular momentum energy as they leave an atom, the remaining angular momentum of the atom must be reduced by exactly that amount carried away. Otherwise the conservation of angular momentum law will not be satisfied.

There has been developed an expression, a simple one, which says that if a beam of circularly polarized light is completely absorbed by the object upon which it falls, an angular momentum:

$$L = \frac{U}{\omega}$$

will be transferred to that absorbing object. U is momentum and (ω) is the angular frequency of the light rays. Thus we find that if (L) is the total angular momentum and (U) is constant, then the smaller the radian frequency (ω) the larger the transfer of angular momentum energy.

It is interesting that in some years past many studies were made, (probably still valid and still being considered) of space ships with "sun sails." These would be large areas of absorbing type materials which would absorb sunlight and thus impart to the space ship a *linear momentum*, which, in turn, is movement through space. It may yet come to pass that some space ships will be the type using some kind of "sun-sails" to traverse the distant cosmic seas!

It is also interesting to us as we consider the phenomena of the atom, which by now we know to be much more complicated then we had earlier suspected; that excitation of the atom causes the production of laser light rays. That there may be unknown side-effects caused by stimulation of the other constituents of the atom may or may not be known as of today. If not known, then when they are determined, we may find some startling new concepts that are far above and beyond the concept of just producing *laser* light.

OPTICAL FIBER FERRULES

We need to return once again to the subject of the first part of this work. We need to examine the optical-fiber ferrule connectors again to keep up with some types currently in fashion. It is said that the Multimate optical fiber connector, shown in Fig. 6-7, is a promising method used today to join optical fiber sections.

Recall that the purpose of an optical fiber connector is to bring the ends of an optical fiber strand, or cable (group of strands) as close together as possible without having them actually touch. Each section, however, must be precisely oriented with respect to its

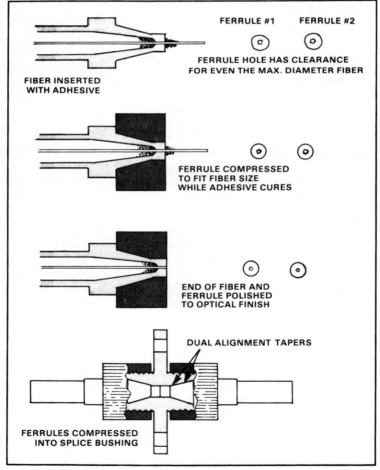

Fig. 6-7. Multimate fiber-optic connectors.

"mate" so that an optically flat surface is presented each way. If the ends are not properly positioned and they happen to touch, then abrasion and wear may take place at the mating point and transmission of light rays is impaired and greatly reduced. Through precise manufacturing methods, it is possible to align the fibers so that tolerances of as small as plus or minus 0.25 mil (in the 6.3 μm range) is accomplished. This happens even though the diameters of the fibers may vary somewhat. The ferrule arrangement shown in Fig. 6-7 makes possible precise alignments and connections for optical fibers. The figure is self explanatory. At the bottom of the figure, the dual alignment taper device permits matching the "flats" of each optical fiber end correctly, and this is necessary to prevent light loss greater than the permitted 2 decibel loss level.

We have been impressed with some small diameter optical-fiber strands we found on a "light-display" type vase in a notions store. They have the multi-strands spreading out from the top of the vase and the ends of these thin strands glow brightly like a myriad of small stars. We took one of these optical fiber strands for a closer examination and found that it is very brittle, would snap if bent very much, and that it is as fine as a hair. We saw the need for the *cladding* (strengthening cover) mentioned earlier in this book. It is important to have strong mechanical protective covering on the optical fibers to prevent breakage when they are put into use in any application wherein they may be moved around such as in a cable arrangement.

Chapter 7
Some Basic Types of Lasers

The uses of lasers is almost beyond imagination. In the medical field alone we find them being used for eye surgery, to "staple" retinas back into place when they become loose, to "weld" the eye lenses into place, and to cauterize wounds. We also find uses of lasers to extend to such mundane subjects as photography where laser color developers are using laser beams to create prints which are said to have near perfect fidelity using 35 mm color slides. In this latter application three laser beams, one red, one green and one blue are used. (Recall the light spectrum? And how, if we use proper synthetic crystals or such we can create light in whatever color we desire?) In any event, these three beams are arranged so that they can scan the color slide or transparency and then, in turn, a receiver positioned on the opposite side gets signals that are proportional to the amount of each light color that passes through the transparency. These electronic signals (that come from opto-electronic sensors that are probably color sensitive) then go to a computer, which controls the emission of three identical lasers that are focused onto a photographic negative. This will cause a print-type negative to be made from the transparency and then color reproductions can easily be made from this print-negative, or "C" print. What is fascinating about this process is that the original art is scanned in such small detail that it is said that as many as 60,000,000 micromeasurements are actually made of the lighting of the original. This produces the detail that is so remarkable! You'll immediately imagine, if you are photographically inclined, that a person might alter the computer program somewhat and thus achieve *special* images, and this is true. It can be done. One type of light emission circuit is shown in Fig. 7-1.

THE SOLID-STATE LASER FOR EXPERIMENTS

Using proper precautions (**you don't want to blind yourself**) you can do some experiments with a solid-state laser, assuming you know where to get one. We'll tell you where you might order one. Try Edmund Scientific Co. 101 E. Gloucester Pike, Barrington, NJ. 08007. They also have optical fibers, lenses, and filters. To "pump" such a laser you might need a circuit using a transistor as shown in Fig. 7-2. This is a suggested circuit, and you will want to monitor the voltage pulses across the laser resistor with a scope. Edmund Scientific may have other circuits which they

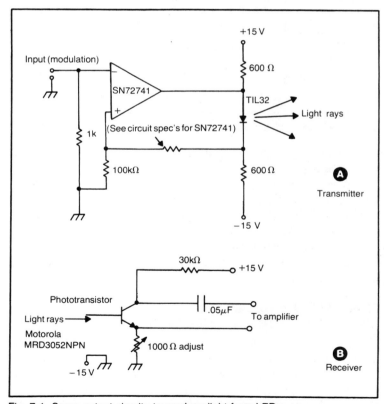

Fig. 7-1. Some output circuits to produce light from LEDs.

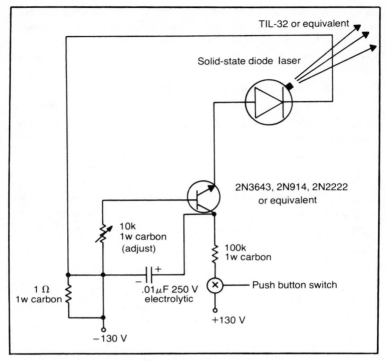

Fig. 7-2. A "pumping" circuit for a solid-state laser or LED.

prefer you use with their laser diodes. **But be careful! Don't look at the light!**

What you have to do with circuit experimentation is to learn what currents, and for how long the pulses of current must exist, or are permitted, for the laser diode you have obtained or plan to get. Once you know this, then you can go about finding a suitable circuit to produce such current pulses. There are lots of circuits available. Just be careful not to burn out your laser with too long a pulse or too high a voltage pulse.

SOME LASER APPLICATION
INSTRUCTIONS FROM II-VI INCORPORATED

A peculiar name for a company: Two-Six Incorporated. They are located at Saxonburg Boulevard, Saxonburg, PA 16056 and they are "into" lasers. They manufacture infrared materials and optics. We know that the infrared part of the spectrum is a good frequency range for laser operation. They have an interest in CO_2 lasers and provide polarization switches that can be used with high-powered pulse generators. It is of interest to us to learn that when utilizing a fast, variable pulse length type of generator, it is now possible to electronically control both the optical turn-on and turn-off performance of a CO_2 type of laser. The switch is called a "Q" switch and it is used with a CdS quarterwave plate. Both are inserted into a CO_2 laser cavity that contains a polarization sensitive element such as a ZnSe Brewster plate. The laser will operate only when the polarization is favored by the Brewster plate. The quarterwave plate inserted in series with the "Q" switch contributes a *round trip* half-way retardation which effectively rotates the preferred polarization some 90 degrees and thus spoils the laser-resonator "Q." This stops the laser operation quite quickly. But when the "Q" switch is pulsed to the quarter-wave voltage level, the unit then contributes a round-trip half-way retardation which nulls out the quarterwave plate component, and that, in turn, *increases* the "Q" of the laser cavity, turning it on. The laser will remain on until the voltage is removed from the voltage-controlled switch. The repetition rate of the switch may be limited by either of two factors. First, the duty cycle restriction of the pulse generator, and by the power handling capability of the Q-switch electro-optic crystal. This crystal is a type CdTe and so its power capability can be increased by water cooling (desirable under some operating situations). A

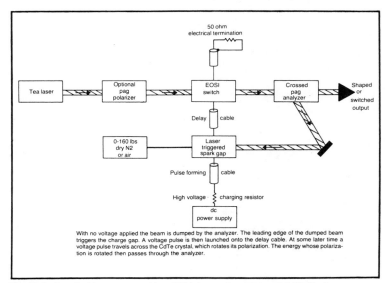

Fig. 7-3. A switching system for a TEA laser (courtesy of II-VI Co.).

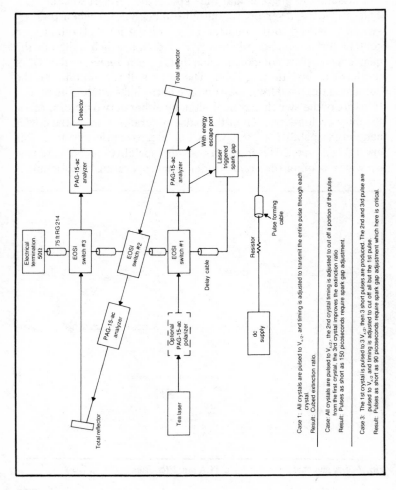

Fig. 7-4. A laser-triggered pulse-switching system (courtesy of II-VI Co).

Case 1: All crystals are pulsed to $V_{1/2}$ and timing is adjusted to transmit the entire pulse through each crystal.
Result: Cubed extinction ratio

Case: All crystals are pulsed to $V_{1/2}$, the 2nd crystal timing is adjusted to cut off a portion of the pulse from the first crystal. The 3rd crystal improves the extinction ratio
Result: Pulses as short as 150 picoseconds require spark gap adjustment.

Case 3: The 1st crystal is pulsed to 3 $V_{1/2}$, then 3 short pulses are produced. The 2nd and 3rd pulse are pulsed to $V_{1/2}$ and timing is adjusted to cut off all but the 1st pulse.
Result: Pulses as short as 90 picoseconds require spark gap adjustment which here is critical.

diagram of a switching system for a TEA laser is shown in Fig. 7-3, and details of the 3 EOSI switch in Fig. 7-4.

HOW DO YOU MODULATE A LASER?

The modulation system changes the phase characteristics of an incident laser beam and then changes the laser's performance based on that change. You pass the laser beam through a unit which is called a "modulator" and control something peculiar to the beam, its phase, frequency, polarization, amplitude, intensity or something in accord with some preplanned intelligent variation. That, essentially is modulation of anything.

In the case of amplitude modulation external to the laser cavity, the incoming vertically polarized light from the laser beam is passed in sequence through a series-arranged quarterwave plate. When no voltage is applied to this plate or crystal the incident light

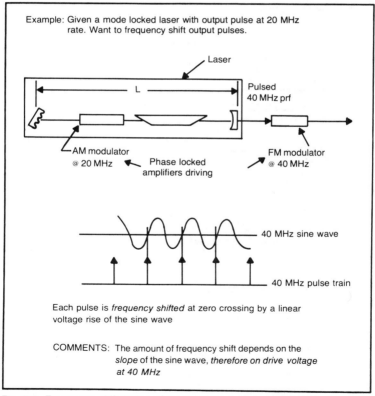

Fig. 7-5. Frequency-shift tuning of laser pulses (courtesy of II-VI Co.).

is converted to, say, a circular polarization and so the amount of light which passes on is reduced to, say, 50% of what its maximum might be. Now, when a modulating voltage is applied to the plate, this causes a change in the polarization to, say, a horizontal value, and if the output elements are such that horizontal polarization is best for transmission, you can readily surmise that the light intensity will increase at the output toward full output value. Also, if the voltage applied is such that it causes a still further shift in polarization toward a direction that is not acceptable to the output elements, then the light diminishes from the circularly polarized (nominal) value and is less in the output. It turns out that practical limitations for linear modulation with this system may be on the order of the 20 to 80 percent range with 50 percent being the nominal value. Modulator units for lasers are available from manufacturers such as II-VI Co. of Saxonburg, PA 16056. In Fig. 7-5 we show a basic schematic of a modulating system for modulating a laser output with an FM signal. In Fig. 7-6 we show some details of frequency tuning a laser.

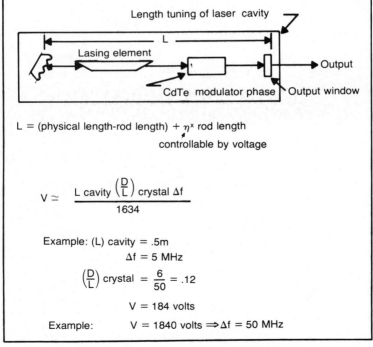

Length tuning of laser cavity

L

Lasing element

CdTe modulator phase

Output

Output window

$L = $ (physical length-rod length) $+ \eta^x$ rod length

controllable by voltage

$$V \simeq \frac{L \text{ cavity} \left(\frac{D}{L}\right) \text{crystal } \Delta f}{1634}$$

Example: (L) cavity $= .5m$

$\Delta f = 5$ MHz

$\left(\frac{D}{L}\right)$ crystal $= \frac{6}{50} = .12$

$V = 184$ volts

Example: $V = 1840$ volts $\Rightarrow \Delta f = 50$ MHz

Fig. 7-6. Frequency tuning of laser (courtesy of II-VI Co.).

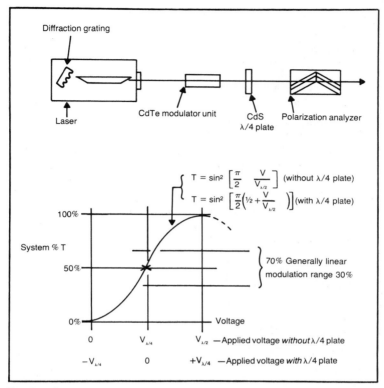

Fig. 7-7. General purpose modulation sytem (courtesy of II-VI Co.).

A general-purpose modulating system for a laser is shown in Fig. 7-7 and some suggested general purpose applications are shown in Fig. 7-8. Finally in Fig. 7-9 we show some additional modulation concepts, and some details of an electro-optic switch in Fig. 7-10.

INTERPRETING SOME CHEMICAL SYMBOLS

In laser and optical fiber discussions we find some abbreviations and chemical designations used that may not be as familiar as pnp and npn. We include some definitions here for whatever benefit they may have when reading this material.

☐ GaAs (gallium arsenide)
☐ Ge (germanium)
☐ CdTe (cadium telluride)
☐ ZnSe (zinc selenide)

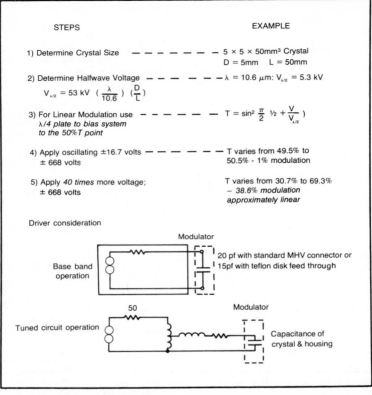

Fig. 7-8. General purpose applications in AM procedure (courtesy of II-VI Co.).

☐ ZnS (zinc sulfide)
☐ GaAlAs (gallium-aluminum-arsenide)

Since there are many types of lenses and mirrors used in this field the following also may be helpful.

☐ Silicon Mirror (Si)
☐ Copper Mirror (Cu)
☐ Molybdenum (Mo)

The types of lasers include the gas-type, which was a new development in 1968. The new development concept was the ability to operate gas lasers at atmospheric pressures, rather than the near vacuum conditions required previously. The laser developed by Lumonics, which has been very successful is the carbon dioxide (CO_2) laser called a TEA laser (TEA Transversely Excited Atmospheric).

224

DOES A LASER BEAM NEED TO BE FOCUSED?

It is normally assumed that you know that a laser beam is focused. It is the focused diameter of the beam and the direction of travel of the beam when used in a cutting application that determines the kerf (the parallelism of the sides of the cut in the workpiece). As you can easily imagine, one might have a beam which comes down into the metal, or whatever, in a kind of slanting direction giving rise to a "V" type of cut (broad at the top and narrow at the bottom). This is not desirable in many applications. With proper focusing and cutting direction, the kerf (or cut) can have parallel sides. They might be from narrow to rather wide depending on the laser's dimensions.

Let's examine some concepts of the laser beam which may be obvious to some but not to others. First of all, recall your childhood experiences with a magnifying glass. To make something "burn"

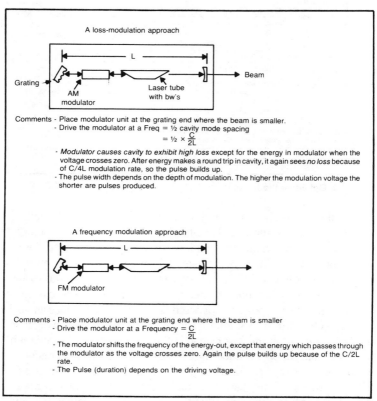

A loss-modulation approach

Grating
AM modulator
Laser tube with bw's
L
Beam

Comments - Place modulator unit at the grating end where the beam is smaller.
- Drive the modulator at a Freq = ½ cavity mode spacing
$$= \frac{1}{2} \times \frac{C}{2L}$$
- *Modulator causes cavity to exhibit high loss* except for the energy in modulator when the voltage crosses zero. After energy makes a round trip in cavity, it again sees *no loss* because of C/4L modulation rate, so the pulse builds up.
- The pulse width depends on the depth of modulation. The higher the modulation voltage the shorter are pulses produced.

A frequency modulation approach

FM modulator
L

Comments - Place modulator unit at the grating end where the beam is smaller
- Drive the modulator at a Frequency = $\frac{C}{2L}$
- The modulator shifts the frequency of the energy-out, except that energy which passes through the modulator as the voltage crosses zero. Again the pulse builds up because of the C/2L rate.
- The Pulse (duration) depends on the driving voltage.

Fig. 7-9. Some added modulation concepts for lasers (courtesy of II-VI Co.).

225

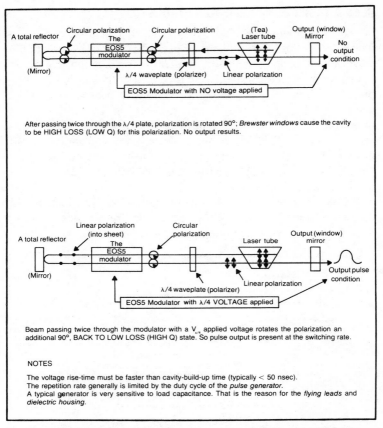

Fig. 7-10. Details of the II-VI electro-optics switch (courtesy of II-VI Co.).

you held the glass so it focused the sun's rays into a very small dot on the object to be ignited. The smaller the dot the hotter the point became. Soon it smouldered and in some cases burst into flame.

A laser beam is normally focused through a special lenses so its concentrated energy is directed into a tiny spot like the sun's rays. The tremendous amount of energy concentrated there can burn through metal or vaporize it or weld it. That focal distance is very sharply defined. For example, if the "spot" is two inches away from the beam outlet on the laser head, then just microns of movement farther or nearer will reduce the energy to the spot. This becomes important in, say, eye surgery, where a laser beam can be focused through the eye to the back wall and there used, to "tack" or weld the retina in place. Be aware that the beam does not damage the eye as it passes through it because it is not focused except at the retina. We now examine in Fig. 7-11 the concept of lens focusing. Of course, there is more to the story than we have stated here. Polarization of the laser beam (Fig. 7-12) plays an important part in

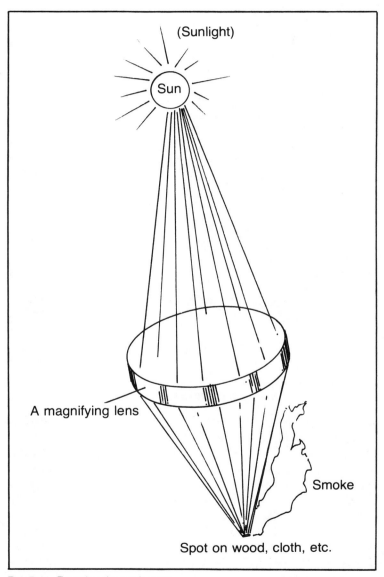

Fig. 7-11. Focusing the sun's rays.

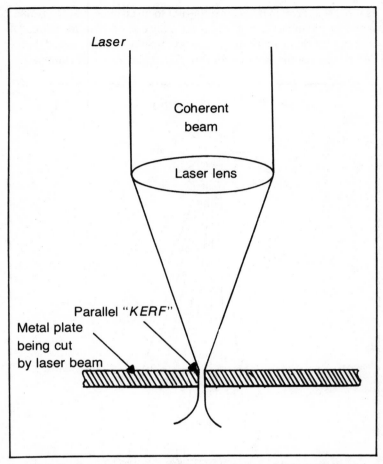

Fig. 7-12. The focus concept applied to a laser beam used in cutting metal.

the focusing concept. Recall that we can control or change polarization with proper quarter-wave reflective phase units.

After the laser beam starts cutting, the metal absorbs the energy from the beam in a parallel-incident manner to the impinging rays and the metal melts or vaporizes. So, what we are doing with the laser beam is to focus it down to a tiny spot, suitable for the task at hand, and then the energy it contains will heat, melt, vaporize or cauterize the object upon which it falls. Of course if less energy is used, either by using less laser power or pulsing the laser less frequently, the energy applied will be less and can thus produce other effects which are useful.

228

HOW IS A GAS LASER CONSTRUCTED AND HOW DOES IT OPERATE?

Basically the gas laser merely circulates a certain type of gas through a sealed cavity chamber by using a pump. The gas passes over two electrodes, positive and negative, to which the high voltage power supply pulse energy is applied. The electrostatic field between these electrodes and the spark causes the atoms of the gas to become "excited" and so they rise to a higher orbit level. Since the gas moves in the chamber, the excitement of the atoms drops as they move away from the electrodes. As they fall back to their normal state, they give off light photons in a lasing beam. This beam is caused to oscillate or travel back and forth in the chamber between mirrors until it reaches a highly sustained and usable level and then it passes out through a suitable "window" in the chamber to be focused by proper lenses and used in whatever manner is desired. Figure 7-13 illustrates in crude form such a laser. Notice the internal mirror system that makes the beam oscillate in the laser head and the window output where the beam is emitted. Of course, the gas (the working element) gets hot and normally it has to be cooled by a heat exchanger. So there are some parts of this system not shown but you can imagine them without too much effort. A suitable working gas is one composed of carbon dioxide, nitrogen, and helium in a proper mixture.

It is important that the plane of polarization of a laser beam be properly oriented when the beam is used in a cutting application. When the plane of polarization is parallel to the direction of travel of the material, or the laser head, the cut has the straightest sides. But, if the direction of movement is at some angle to the plane of polarization, a wide cut results with slanted sides. If you still increase the angle, when it approaches 90 degrees from the polarization plane, the cut will be wide, ragged, and have rounded sides.

We have learned that the ruby laser operates in the infrared part of the light spectrum. Now we need to know that the gas lasers we are considering can operate in the infrared and also the ultraviolet portion of the light spectrum. All types of lasers can produce coherent light and are able to concentrate the energy contained in their beams in a small area to create high temperatures.

OTHER LASER TYPES USING GASSES

There are other laser types. These range from the rare gas halide (excimer) laser developed by Lumonics in 1978 to the mini-xenon chloride laser (a nitrogen type of laser) the silicon marker

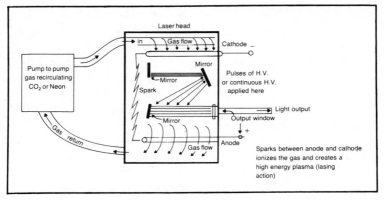

Fig. 7-13. A basic sketch of a gas-type laser.

laser (can mark silicon semiconductor wafers) to the dye laser that uses tunable light in the visible spectrum. It is also of interest to know that in some cases an ultraviolet flash source using its own electrical supply system may be used to pre-ionize a discharge cavity prior to the time that the main spark pulse is delivered to the cathode and anode in the lasing chamber. (Neon-helium, and carbon dioxide mixtures have been very successful in gas-types of laser instruments.)

WHAT IS A PYRO DETECTOR?

A pyro detector is broad-band and designed to make pulsed energy measurements of laser systems of various kinds. They are capable of handling high energy pulses without damage. The detector element is constructed from a high Curie-point material and this is protected by a ruggedized case. The detector will deliver a voltage pulse whose amplitude is proportional to the number of joules impinging on its surface. The output is said to be independent of beam width or its pulse length.

OPERATION OF THE LUMONICS TEA-820

As suggested in Figs. 7-13 and 7-14, such a laser is based on a pair of transverse-profiled electrodes providing a spark volume discharge of greater than 300 cubic centimeters, housed in a 25 centimeter fiberglass vessel. A rigid structure of rods hold the optical components inside the vessel. Lasing gas (a gas that is known to be capable of population inversion excitement and relaxation and thus production of a laser light output) is circulated transversely over the electrodes. While this is happening, electrical energy from a charged capacitor is periodically discharged through the spark gap into the gas. That produces the population inversion and the laser action that produces the laser light beam.

It is appropriate here to define the excimer laser. This type is very much like the TEA laser except that it produces light in the ultraviolet portion of the spectrum instead of in the infrared portion. It is also good at this time to get some idea of the pulsing energy needed to operate such a laser. In the TEA 600A, CO_2 gas laser, a charging voltage of up to 50 kV is needed. The operation of some of these type lasers requires as much as 800 to 1000 joules of electrical energy to be delivered into the discharge cavity where the gas is located, during a one microsecond pulse. This can require from 140 to 160 kV of electrical energy and a maximum of inductance in the energy storage network. It reminds one of the pulsing networks used with radar systems doesn't it? It is the same idea and same pulsing principle. In a rare-gas type of laser, halogen gas and helium are mixed and supplied to the laser cavity. Another rare type gas is xenon. Laser cavities are usually pumped with a vacuum pump to clear out discharged gasses.

Finally, we mention that in case you'd like a laser kit to experiment with, you can write to: Lumonics at 105 Schneider road,

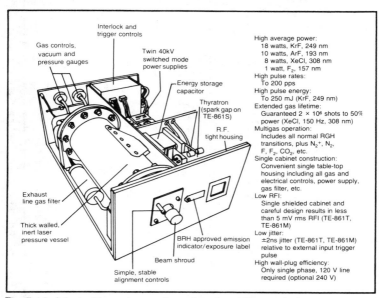

Fig. 7-14. A typical lumonics gas laser (courtesy of Lumonics).

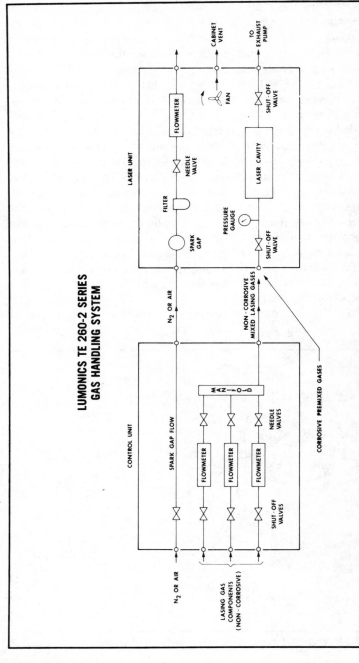

**LUMONICS TE 260-2 SERIES
GAS HANDLING SYSTEM**

Fig. 7-15. Engineering diagram of Lumonics gas-flow to laser (courtesy of Lumonics).

Kanata (Ottawa) Ontario, K2k 1Y3. In Fig. 7-15 we show the gas-handling details of their TE 262 laser.

In Fig. 7-16 we show the schematic of a dye laser. Notice that an excimer pump (ultraviolet output) laser is used to excite this dye laser. Also, note that a diffraction grating is used in the tuning mechanism. The excimer pumping beam goes to the dye cell oscillator (3) where the laser beam is formed. The beam then goes into the tuning mechanism for reinforcement by diffraction grating adjustment at near grazing incidence. A low magnification (15×) four prism achromatic beam expander is used to get high conversion efficiency and narrow and constant line width on the output beam which then passes to the output coupler (4). Notice that what happens here is a positive type of feedback at the desired frequency through the dye cell reinforcing the energy until the beam is of sufficient intensity to be used. It is then coupled through an amplifier dye cell. The amplifier dye cell also is pumped (maser action) by the excimer beam.

OPERATION AND DETAILS OF LASER WAVEPLATES

Since the laser outputs are light rays, we are concerned with focusing of these rays, the polarization of the rays, and the fre-

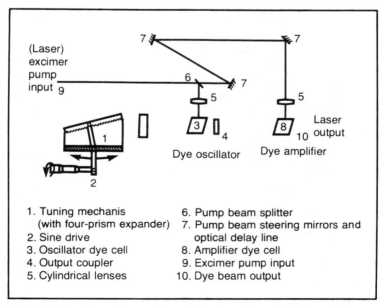

1. Tuning mechanis (with four-prism expander)
2. Sine drive
3. Oscillator dye cell
4. Output coupler
5. Cylindrical lenses
6. Pump beam splitter
7. Pump beam steering mirrors and optical delay line
8. Amplifier dye cell
9. Excimer pump input
10. Dye beam output

Fig. 7-16. The schematic operation of a dye laser (courtesy of Lumonics).

233

quency or wavelength, and reflection of these rays. In fact we are concerned with all elements which enable us to control, manipulate, direct, and use the light rays to our advantage. One such element is the *waveplate*.

When a plane polarized beam is incident on a birefringent plate whose *optic axis* lies in the plane of the plate, the beam is resolved into two components. These components propagate through the plate with different velocities and recombine upon exiting. There they have a new and different polarization. The character of this polarization will depend on the orientation of the optic axis and also on the net difference in optical path length of each of the two components. This will govern the phase relationships of the two components of the wave and thus the resultant phase of the combined wave. This difference in optical path length (d) is given by an equation:

$$(d) = (\Delta n)\ t$$

where (Δn) is the birefringence of the material.
(t) is the thickness of the material.

When (d) is equal to one quarter wavelength of the frequency being used in the light ray, and *the optical axis* of the material is at 45 degrees to the incoming polarization, the outgoing polarization of the ray is circular. This is needed and necessary in some types of applications. But, when (d) is equal to one-half wavelength of the frequency being used, then the output polarization is linear, but it is rotated with respect to the input polarization by twice the angle between the input polarization and the optical axis of the material in the waveplate. Notice we are careful to state the optic axis of the material. This may or may not coincide with the physical axis of the material.

The thickness of quarterwave plates made of CdS for a frequency of 10.6 μm is very thin, making it fragile and hard to manufacture (.1 inches thick). Such plates are made by II-VI Inc. to be odd multiples of that quarter wavelength (3/4, 5/4 or 7/4 wavelengths for strength). The plates are polished flat to better than $\frac{\lambda}{40}$ at the frequency of 10.6 μm, and are parallel to within two seconds of arc. They have an antireflection coating applied at the designed wavelength. Some of these waveplates have a transmissibility factor of as high as 97 percent.

If you "retard" a light ray (as is often mentioned in the current literature) you actually change the phase or polarization of that light ray. Often it is necessary to have some variable phase controls over the light ray and for this purpose an element called the Babinet-Soleil Compensator is used. This is constructed of two wedge-shaped plates, oppositely oriented, and this produces a variable multiple order plate. If another fixed thickness produces a variable multiple order plate. If another fixed thickness parallel plate with its optic axis rotated 90 degrees is added to the two, then one has a variable equivalent order zero plate.

The compensator consists of two CdS wedges with the wedge angles in the opposite directions. One such wedge is adjusted by a micrometer in a tranlatory manner in the direction of the wedge. Thus the two wedges serve as a variable thickness piece whose birefringence is offset by an added parallel plate, so the net effect is to produce a variable retardation, adjustable by the micrometer setting. Such a device is adaptable to beam splitting types of operations.

BEAM SPLITTING AND LASER-RANGING MECHANICS

We know that the separation of various polarizations of the same wavelengths of a laser beam is often accomplished by using a Brewster window (see Fig. 7-17). We see that essentially one half

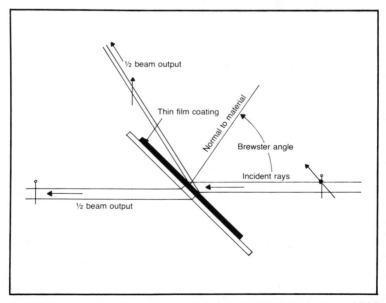

Fig. 7-17. A TFP (thin film polarizer) beam splitting concept (courtesy of II-VI Co.).

of the incident beam is bounced off the reflective surface with a polarization change, while a second half of the incident beam goes through the material with essentially the same polarization as the incident beam. The incident beam is thus "split" into two components whose phase is somewhat different. If the rays are later recombined, the cancellation or reinforcement effect can be measured and used constructively.

In the field of laser ranging, such a device can be used in the following manner: A "P" polarized pulse is sent out through the beam splitter, though a phase adjuster (retarder) to convert the linear polarization to a circular polarization and then the laser ray goes on to impact the target. Assuming reflectivity, the ray is bounced off the target and returns, again through the wave (phase) adjuster so that it is converted into a linearly polarized wave whose phase is slightly different from the incident ray. Upon impact with the beam splitting element, this time on the thin film coated side of the element, it is reflected away at some angle to a detection unit. Its travel time can be measured and thus the range to the target electronically calculated and displayed through suitable light-emitting diode units, if that is desirable. The so called element used in this application is called a *Brewster plate*. It can have a transmission of light rays through it when its noncoated side is presented, at some angle, to the incident beam. It can cause *a reflection* away from itself through some other angle, of an incident ray impinging upon its coated surface. The company II-VI of Saxonburg, Pa. 16056, makes such units and has more details. Figure 7-18 shows elemen-

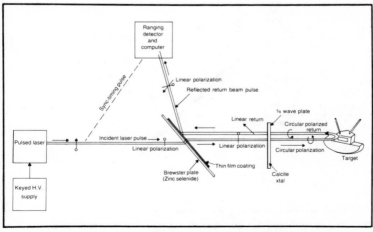

Fig. 7-18. A laser ranging system concept (courtesy of II-VI Co.).

tary details of the use of the Brewster plate in a laser ranging concept. This might be a military device or in a surveyor's instrument, or in some other type application where the distance to some reflective object is, or can be, determined through the use of laser pulses.

It turns out that in applications of such as a calcite crystal that has been cut to a quarter wave thickness at some light frequency, and then the ray passed through it, the emerging ray will be circularly polarized, as illustrated. It also turns out that the optic axis of such a crystal is not through the physical axis precisely. One may construct a line through a corner of the crystal at some precise angles to the faces and this line then is the optic axis. Once determined, however, a crystal may further be machined so that the optic axis may coincide with a physical axis if this is desired.

POLARIZATION

Polarization is said to be the direction of the $(\hat{E})$ electrostatic vector in a wave composed of electrostatic lines of force and $(\hat{H})$, the electromagnetic lines of force. The polarization of the wave described by the direction of the $(\hat{E})$ vector may be varied. The polarization may be parallel, or circular, or perpendicular to the direction of the light ray. *Faraday rotation* of an electromagnetic wave means that upon impact with a reflecting material, the $\hat{E}$ and $\hat{H}$ vectors rotate in opposite directions to each other and to the direction they had upon incidence with the material.

When describing the phenomena of refraction, reflection, and transmissibility of laser rays through or from surfaces of various materials, it is customary to indicate what happens to the polarization of the ray as it passes through or is refracted or reflected from such a surface. Since we can assume that polarization is related to phase, then, since we know how phase combinations affect the amplitude of a resultant wave and the phase of a resultant wave, we gain some idea of what happens in laser light rays if such rays are divided in any manner and then recombined. Suppose, for example you have a beam splitter that accurately divides the power in a laser beam exactly in half. Suppose, further, that the polarization between the two resultant beams is changed by 180 degrees and then the beams are recombined. Intuitively we would say that the two beams cease to exist since they cancel. This is the same idea used with phase relationships in any type electrical-electronic circuit and radio-radar, or microwaves, etc.

When we discuss light rays and their use and control we get

quickly into the subject of polarization. Some definitions therefore amplify what has previously been written.

Linearly Polarized Light. That condition where the electric (ê) vector, describing the electric portion of the wave, vibrates in one direction only and that direction is perpendicular to the direction of propagation of the wave.

Circularly Polarized Light. The direction of vibration of the wave rotates around the direction of propagation in a corkscrew fashion. No single direction of vibration is favored. Be aware that the direction of rotation of a circularly polarized wave can be in either direction—right-handed or left-handed. This refers to the concept of vectors where, if the fingers of the hand curl in the direction of rotation, the thumb points in the direction of propagation.

Elliptically Polarized Light. This is not a concept we are usually familiar with. This can be considered to be somewhat the combination of both linear and circular polarization, where the rotation of the wave is favored slightly more in one direction making the resultant looks somewhat "egg shaped" or elliptical.

Unpolarized Light. This means light rays which have some kind of changing polarization in some unpredictable manner and on a random basis. No particular direction of polarization is favored. Unpolarized light can be segregated so that a particular polarization can be favored by passing the light through a "polarizer" or using a diffraction grating.

THE POYNTING VECTOR

This vector addition shows the propagation direction of an electromagnetic radiation. Since we are considering light to be in the extended spectrum of electromagnetism, it is well for us to examine the Poynting vector shown in Fig. 7-19. We know that the electric field and the magnetic field of a propagated wave are 90 degrees displaced from one another. Thus, if we show two arrows displaced by 90 degrees, whose length represents the magnitude of the wave component, and whose direction represents, instantaneously, the direction of those waves, then the vector addition of the two represents the total or the Poynting vector. You "curl" the $(\hat{E})$ vector into the $(\hat{H})$ electromagnetic vector and the thumb points in the direction of the resultant $(\hat{P})$ vector. The magnitude of the resultant is the absolute value numerically of E and H multiplied together times the sine of the angle between them. When this angle is 90 degrees the sine is 1.

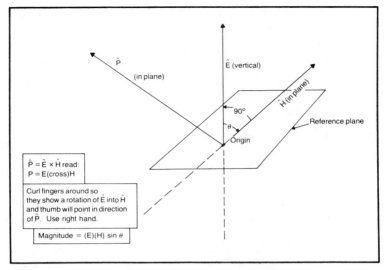

Fig. 7-19. Method of vector multiplication on the poynting vector.

From the above we can clearly see how a light beam ($\hat{P}$) might be decomposed into its two orthogonal (right angle) elements ($\hat{E}$) and ($\hat{H}$) whose amplitudes are such that with vector addition they can be recombined into the original beam. It is sometimes said that a beam of unpolarized light can be represented as two right-angle components of linearly polarized light which are equal in amplitude.

THE EFFECT OF LIGHT POLARIZERS

These devices tend to pass light components that are polarized in the same manner as the "polarizer." They can be made to act as light switches. If we take two of them, they are like sunglasses that pass only a given polarization through them. If we rotate them so they both pass the light rays of a given polarization (linear) then maximum transmission of the light beam results. If we rotate one of them so that its direction of polarization is 90 degrees away from the first, then no light will be passed. In a sense, we have a light switch, and since a laser beam is monochromatic and is polarized light, we can use such a switch very effectively in some applications where laser beam control is useful.

We must note that nothing is perfect. There is not a perfect extinction of the light transmission even when the polarizers are properly adjusted. The rate of light transmission when the polarizers pass the light, to the rate of light transmission when the polarizers are adjusted to cut-off the light transmission has a limit. It is called the "extinction ratio" of the polarizers.

METHODS OF POLARIZING LIGHT

The laser generates the "in-phase" or polarized light ray. Light can also be polarized by reflection at other than some normal incidence from an interface of two dielectric materials of different indices of refraction, such as that between air and glass. If the angle the light hits the reflecting surface is equal to the Brewster angle (equal to that angle whose tangent is $\frac{n_1}{n_2}$, then the reflected component of the beam is entirely polarized *perpendicular* to the plane of incidence. The part of the beam that penetrates the interface is small and is partly polarized parallel to the incidence plane. By considering two materials which have different indices of refraction, and passing a light through them at a proper angle, we can get a perpendicularly polarized component of the light beam or a parallel component of the light beam. This also is useful in some applications.

Notice how we can control the "flow" of a light beam by having it polarized and using polarizers in a proper manner. The same idea is true with electromagnetic radiations in radar and microwave applications. Ferromagnetic devices can control the polarization of radar and microwaves in much the same manner that some types of crystals, glass, and other substances control the polarization in light beams. There is a device known as a "wire grid" that can polarize the longer wavelengths in the infrared spectrum, if the spacing of the grids is small compared to the wavelength.

It is possible to absorb one component of a light beam that has some given polarization. This is called *polarization by dichroism.* The phenomena is accomplished by orienting a long chain of molecules in one direction so light that is vibrating parallel to them will pass through the molecular material. Light that vibrates at right angles to the direction of the chain (orthogonal) is absorbed and does not pass through the material. Oriel Corp. makes sheet polarizers made of an absorption of iodine in a sheet of thin polyvinyl alcohol that has been stretched so it causes the molecules to be stretched in a long chain. It is a "polarizer" for light beams of given frequencies.

To further understand the control of light rays, we again refer to the calcite crystal. This is a parellelpiped type of crystal with a well defined *optic axis* as illustrated in Fig. 7-20. As you note from the figure, there are many parallels to the optic axis, which is a direction through the crystal. This crystal is peculiar for it will show a *double refraction* (two beams emerging from the crystal) if the incident beam is *perpendicular* to the optic axis. If the incident beam's direction is parallel to the optic axis, there is no double refraction.

Now we come to another definition which is important in optics and light control. We call one part of the beam which passes through the calcite crystal the *ordinary* or (O) ray. The second component of the incident beam we call the *extraordinary* ray or (E) ray. The ordinary ray is parallel to the direction of travel, in its polarization. The extraordinary ray is polarized perpendicular to the direction the beam is traveling. Thus the polarization of the ray component of the beam determines whether the ray will be called an ordinary ray or an extraordinary ray, and have the abbreviation O or E.

So, what's the importance? The importance is that when a beam having these two polarization components (and ordinary light has them) passes through a calcite crystal, the beam is essentially split because the index of refraction of that crystal is different polarizations of the two rays, or beam components. This index of refraction is said to vary from 1.66 for an E ray to 1.49 depending upon whether that E ray component travels parallel to the optic axis or perpendicular to it. By shifting the physical orientation of such a crystal with respect to a beam of, say, laser light, the refraction of

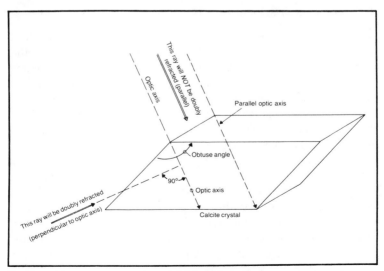

Fig. 7-20. The optical axis of a calcite crystal.

that component of the beam can be shifted as the index of refraction is shifted.

If an unpolarized beam is passed through a calcite crystal, it has been found that the light will become polarized and have two components perpendicular to the crystal's optic axis. One will be the O ray and the other our familiar E ray. Now here is a fascinating part of this study. Both rays will travel through the crystal at the same speed but they will be refracted differently and so the difference will become evident. The O ray normally will be refracted according to Snell's law of optics. But, the E ray will change its direction somewhat, due to the effect of a different index of refraction and it will move away from the optic axis and out of the plane of incidence. The emergence of the two rays from the crystal will be parallel to the incident beam direction but both will be shifted somewhat due to the effect of the crystal and they will now emerge in parallel lines. In effect the O ray will separate from the E ray and we will have two light beams emerging where we had just one incident beam. Fascinating, isn't it? As we have stated earlier, when a calcite crystal is placed on a page of a book like this, and the lines viewed through it, the lines and letters will appear doubled, or to state it another way, you'll see two sets of lines and characters. This phenomena is called *double refraction* and it has its place and usefulness in handling and control of light rays. A final thought: Some scientists believe that the speed of travel of the two rays may be different and if the index of refraction is defined as $\frac{c}{v}$ where c is the velocity of light and v is the actual beam speed, then there is a variance in the index of refraction for one ray with respect to the other, and so more bending of one ray than the other takes place.

There are several types of doubly refracting crystals such as ice, quartz, siderite, calcite, dolomite, and wurzite, to name a few. Each will exhibit this ray splitting phenomena when a beam of light at about 5890 angstroms is used as the incident beam. Notice here that *the frequency* of the incident beam may be of importance. What may be a phenomena and effect with a beam of one frequency and wavelength, *may not* appear with a different light beam of a different wavelength and frequency. Also, one assumes a nonpolarized incident beam. Just what the effect of passing various frequency laser beams through these type crystals is, we leave to your scientific curiosity and experimentation. The laser beams, remember, are polarized. Thus the orientation of the crystal with respect to a laser beam may produce various effects.

WHAT IS A POLARIZING PRISM?

Still concerned with polarization, we examine the types of prisms which might use the polarization effect of the light rays. A *polarizing prism* is as good as its name. It polarizes the ray passing through it. If we want to polarize some incident light we can use a polarizing prism to do the job. It will pass the E ray and restrict passage of the O ray and the result will be a beam of polarized light. The prism, one type at least, is made of two wedges that are positioned so that there is a tiny air space between them (Fig. 7-21).

If you focus some unpolarized light on the element so that it enters the crystal prism perpendicular to the optic axis, the (E) ray will go through the first prism, hit the air space and then go through the second correcting prism and emerge essentially unchanged. The O ray will "take-off" from the prism-air interface at some wide angle and "get lost" somewhere else. Thus the two components have been separated and we find ourselves left with just the *polarized* (E) beam coming through the element. That was what we wanted in the first place, for some unknown reason. The prisms can be made from calcite.

There is a critical acceptance entry angle for the incident beam. If too low an angle is used, it will be *reflected*, if too great an angle is used, some O ray gets into the desired output and "messes-up" our nice linearly polarized E ray. The acceptance angle for some prisms of this type is around 9 degrees. Oriel Co.

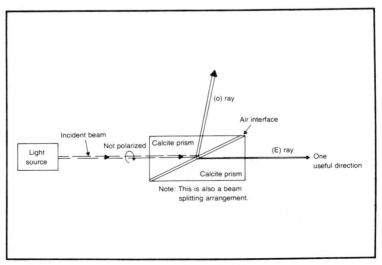

Fig. 7-21. A polarizing prism passes the (E) ray.

makes such a prism polarizer with *extinction ratios* of up to 10^5 and that is good.

We need to be aware that polarization can be used to cause certain phase relationships to exist in a light-ray-wave. If elements with optic axes are used to pass the beams, then we find that there are, essentially, a *fast* and *slow* axis of propagation in these crystals. This is governed by how the polarization of the light ray is oriented with respect to the optic axis of the crystal. A ray polarized parallel to the optic axis is said to be the "slow" ray, while the one that may be polarized perpendicular to the optic axis is said to be the "fast" ray (E ray). Thus, as we have indicated earlier, there can be specific phase relationships, or polarizations of the emerging rays, depending on time and distance of the wave passing through the crystals, and the wavelength-distance of the crystal element.

Retardation means to an electronics buff that the phase of the light ray has been caused to "lag" the fast axis propagation. Also, these type plates can produce, with proper polarization changes, a circular polarization from a linearly polarized beam. This is important in many applications. Recall that a laser produces linear polarization of its beam. To convert this to circular polarization, as required for ranging, for example, one uses retardation plates, which are crystals through which the beam must pass.

EXPLAINING POLARIZATION SYMBOLS

When drawing diagrams showing light beams and where they go and what happens to them, it is useful to use some symbology to indicate the polarization of rays or beams as they move through various elements of light control. Such symbology is shown in Fig. 7-22. First, notice that the ray or beam of light is represented by an arrow with an arrowhead (vector) that shows the relative direction of propagation of the light ray or light beam. Second, notice that the polarization is indicated by drawing short lines or circles or combinations of short lines and circles (C), or just partial circles (rotating vectors) on the ray or beam lines.

By using this type symbology it is easy to indicate how the polarization of rays or beams of light may change when going through the various blocks of a system of light control. We need to say that in Fig. 7-22A, the small cross-lines may be given arrowheads so they become small vectors to indicate whether the polarization is up or down as we see it graphically. This could be, say, 90 degrees or 270 degrees depending on the cross-arrow direction. Also, if the ray or beam vectors were realistic they not

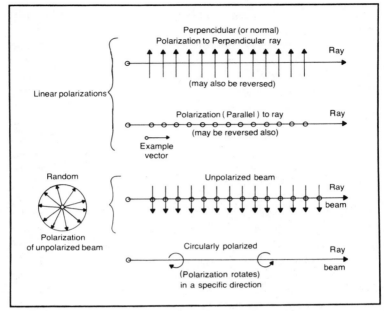

Fig. 7-22. Representations of light-beam polarization symbology.

only would show by their direction, the direction the light is propagating, but also, indicate by their length *the relative* intensity or strength of the ray or beam. Sometimes the length aspect of the vector concept is *not* useful, so be careful.

USE OF PRISMS

There are many types of prisms and these can be used to control and direct the light of a laser beam. We will list some of these for your information.

The *corner cube prism* is so designed by the manufacturer that it will return a beam of light back toward its origin after three internal reflections regardless of the orientation of the prism with respect to the beam of light. A *90-degree reflector prism* is a right-angle wedge which will deflect the light rays or beam 90 degrees from the incoming direction. The so-called *roof prism* is like the 90 degree prism in that it is a wedge shaped section of, say, silica glass. Actually it is an equilateral triangle and if the light beam enters the base, to one side, it will be reflected from the two sloping sides and come back out the base some distance away from the entrance point. Thus this type prism reflects a ray or beam back toward its source but unlike the corner cube prism the ray is spaced from the entrance line. It is not reflected back on itself, precisely.

A *laser dispersing prism* is one made from fused silica glass and polished, and the laser beam is adjusted so it passes through at the Brewster angle and thus provides small loss with polarized light inputs. The beam goes into one side of the equilateral triangle, across the material, and out the other slanted side. The beam is "spread" out, as some different frequencies may be bent more than others. If a given laser has just one frequency component, it is bent just a given amount (refracted) but if several monochromatic laser beams impinge on the prism (and each is different in color or frequency from the others) the output will be a separation of those frequencies by physical spacing.

If a *dispersing prism* is cut in half then the result is called a *Littrow prism*. If the desired beam enters the hypotenuse, it will reflect from the flat rear side, and will come back out in the same direction it entered. If the input beam happens to be from a *multi-frequency laser* (there are such things) the various rays separate and you can select the frequency of beam or light ray you desire quite easily. The prism is made from *Schlieren* quality fused glass.

Finally there is a *penta-prism* that will deflect a given beam 90 degrees no matter how it enters into the prism. This Penta-Prism is said to be useful in alignment of lasers and of auto-collimating systems. It is possible to rotate this type of prism and so create a "plane of laser light" that has some use in leveling operations and such. We recall the construction of a room for a large commercial computer. The floor had to be exactly level. To accomplish this, a device was used that had a gyroscope that was self-orienting if left running a few minutes prior to use. When it leveled, it adjusted a laser that then flashed a light beam around the room, clearly marking a line on the walls. The floor was then measured down precisely from this line and, believe me, that was a level flooring! In a sense this was a laser-light plane throughout the room used as reference. We must mention the *beam splitting cube* that consists of two prisms separated by a thin air interface, as previously mentioned, and the *dove prism* that rotates an image about the beam axis by an angle equal to twice the angle of the prism as measured to the beam. If you are interested in these or other special types of prisms Oriel, at 15 Market Street, Stanford, Conn. 06902 is a good source.

A DIAMOND AS A WINDOW?

A diamond will provide an excellent window in a wide-band

sense from about 0.23 μm to about 200 μm (micrometer) in the infrared band. The diamond is hard, durable, has high tensile strength, low thermal expansion, and high thermal conductivity, so they are excellent for *high powered* windows for lasers. The diamonds used for this purpose are called the type II diamonds. Of course other types of "windows" can be used but the diamond is an interesting one isn't it? Zinc selenide and germanium are some of the other types. Of course the diamond or the other elements are ground into flat plates of very thin dimensions and then can be used to pass the beams of light out of chambers, tubes, or whatever.

FILTERS FOR LASER LIGHT BEAMS

There are many types of filters that pass only certain amounts of light or pass just certain frequencies of light. These are important in light-control. The filter operates either by attenuating the light ray or beam or by absorption of that light ray or beam. It is possible to attenuate a light beam by passing it through an element that causes *destructive* phase relationships (polarizations) to exist that cause elimination of the rays.

A reflective type of neutral filter can consist of a thin metallic coating on a clear substrate or base material. The amount of light passing through the coating depends on its thickness, as you would intuitively guess. Light that is not transmitted through the filter is partially absorbed and partially reflected. In any event it does not get through the filter. The purpose of the filter is to cause certain bands, frequencies, or colors of light to be eliminated or not pass on to other elements. Or it can be used to reinforce certain frequencies or colors. The substrate material is usually glass or UV-grade fused silica. An absorption type substrate is usually a grey-colored glass that absorbs the light within itself. The kind and type of glass and its thickness determines the amount of absorption.

Neutral density filters are used to fill a need for optical light attenuators as used in photomultipliers, radiometers, thermopiles and such instruments. These kinds of instruments are usually sensitive to many frequencies of light and it is desired that the responses be uniform over broad bands. The types of filters used help to smooth out the peaks and valleys of various wavelengths without omitting any of the wavelengths of the incident light.

Sometimes various types of filters are used, as they are in good photography. By the way, a study of camera photography and filters will help you to understand their types and uses in light and color (frequency) control of light rays and beams. It is easy to mount such filters on a color wheel. Turning the wheel places various filters in the beam path and the effect of the filter can be studied. The content of the beam's spectrum may be easily analyzed in this manner.

NARROW BAND FILTERS

These can be likened to the filters used in radio receivers. They pass only a very narrow band of frequencies of the light beams and attenuate all others. They are like the i-f system in a superheterodyne in the way they work with light. The basis of this type filter is a series of thin-film partially-reflecting layers arranged as a single or multiple Fabry-Perot interferometer. This interferometer will cause destructive annihilation of undesired light rays within itself. We suggest reading about Michelson's interferometer in a good college physics text.

In any event, considering the multiple layer interferometer, the spacings between layers is chosen so that beams produced by the multiple reflections from these layers are in-phase with the transmitted wavelength for the desired frequency, thus they go through the filter quite easily. Other wavelengths are not transmitted because of the destructive interference just described. By varying the reflectivities, the number of layers, and the number of cavities, in which the filters are placed (these are called Fabry-Perot cavities) a wide variety of wavelengths and beam shapes can be produced. The center wavelength is always determined by the thickness of the thin film deposit.

FILTER TERMS AND DEFINITIONS

Light filters are described by various terms.

☐ Peak wavelength—The wavelength of maximum transmission.

☐ Central wavelength—Frequency half-way between half-power transmission values.

☐ Peak transmission—A maximum transmission value.

☐ Bandwidth—Width in wavelengths of the maximum value obtained between ½ levels of maximum value. (Essentially the ½ dB value as we in electronics know it.)

☐ Band-Pass Shape Factor—A graph of the band-pass characteristics of the filter or filters.

☐ Blocking Region—Just as it says, the wavelengths which are blocked by the filter.

☐ Incidence Angle Shift Factor—Also called the effective index-of-refraction. Gives the change in peak wavelength with a change in filter angle of incidence. As incidence angle is increased, the peak wavelength becomes shorter. Changing the angle of incidence is also called "tuning" a filter.

There are many types and kinds of filters. This discussion is somewhat limited, but will provide you with enough background so that other types may be readily understood, and their uses evaluated.

THE ETALON INTERFERENCE FILTER

You'll see this word if you read about the use of filters. It is an important device in laser light control. It is a Fabry-Perot interferometer. It can be used to narrow the bandwidth of a transmitted beam (from, say, a dye laser, which generates a relatively broad band of wavelengths) by causing the undesired ones to interfere with one another *destructively* and the desired wavelengths to combine *constructively* within its reflective boundaries. When the Fabry-Perot interferometer is placed within a cavity that is tuned to some exact wavelength—so that a given beam frequency is produced, the interferometer is then called an Etalon.

LIGHT-REFLECTING MIRRORS FOR LASER BEAMS

In some types of lasers a "cavity" is used to resonate at light frequencies. This cavity is like a tuned circuit in its operation. (Good books on RADAR will enhance this explanation.) We are concerned with the transmittance of energy inside and through such devices. With light frequencies, mirrors are used to cause reflections in laser cavities so that energy is imparted to solid-state elements such as rubies, or gasses such as helium, or dioxide mixtures, and dyes, which absorb and release energy in the form of light photons to cause lasing beams of light at certain frequencies. The cavity, which is a tuned circuit resonates only at one frequency and is relatively narrow band, thus it helps to enhance the creation and transmission—out of itself—of a specific laser frequency of light which is monochromatic and has a single polarization.

Some mirrors are designed to absorb certain light wavelengths and pass other light wavelengths. Thus, they may act as filters. For example some mirrors will reflect heat (infrared) and pass visible light. Others will reflect visible light and pass infrared. These latter types may look like a metallic mirror, but they may be coatings on highly reflective pyrex glass that is heat resistant. Mirrors as used with lasers may be called *reflectors* and may be elements that are polished on only one side to a very high degree of flatness and surface polish. They may be uncoated substrates with a coating of silicon monoxide or aluminum, and the flatness (measured in wavelengths) may be as "tight" as 1/20 of one wavelength.

There is a mirror called a *first surface concave reflector*. These are ground and polished, highly sperical surfaces and they also may have the coatings just described on a type of substrate. The concave side of the mirror is highly polished, while the "back side" may be flat and is normally ground. The edges are bevelled. They have a given focal length, concentrating the rays impinging on the concave surface back to a point some number of wavelengths away, and somewhat parallel to the arrival axis. The size may be from two inches to six inches depending on the application.

The laser cavity end-mirrors and interferometer mirrors are made from a substrate of polished glass or silica that has appropriate reflector coatings of film applied. The rear reflector normally has a single maximum reflective coating, and the front reflector has a partially reflective coating. On the outside an anti-reflective coating is applied so laser light rays won't get back inside the cavity. This latter type of mirror is, in a sense, a window, for when the laser beam reaches a given density or strength it will burst out through this "mirror" and beam to where we desire it. The mirror surface on the inside is required so that the laser beam, when building up, will bounce back and forth inside the cavity until it becomes strong enough to emerge. Figure 7-23 shows this concept.

Some mirrors for high-powered lasers of the CO_2 types (we

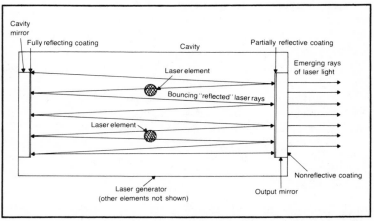

Fig. 7-23. Mirrors are used in a laser cavity to bounce rays back and forth to stimulate emission.

define a "high-powered laser" as being one which can produce ten watts or more of energy output) may be concave or flat and it may have its coating surface of gold-copper, or "hard gold." The base may be occluded nickel, which is coated on copper. This material has a very high heat resistance and a very high thermal conductivity. The mirrors are total reflectors and can be used up to the multi-megawatt region of power outputs.

If you can use gold in one way you can use it in another. So why not coat a pyrex substrate with gold? Oriel Co. does this and produces a low-power type of mirror for low-powered laser systems. These mirrors are polished to within 1/80 wavelength and the gold coating gives a high degree of reflectivity in the infrared region.

This brings up the concept of "roughness." We think of polishing something as simply removing black splotches or some undesired coloration, or, in a wood-working sense, we think of smoothing down a wood surface until we can feel no grain by hand. Well, it is of interest to us that if we go out into space some millions of miles, the earths surface looks absolutely smooth. We know it is not because we live here. So, if we look at a surface under a given magnification of microscope, we find that it too may not be "perfectly smooth." Did you know that actually the earth's surface is "as smooth as a bowling ball? It is, and that tells you something about the surface of a bowling ball. Think of it the next time you run your hand over such a surface. This tells you something about the smoothness of laser mirrors, which may be polished to within 1/80 of a wavelength of light. Think of the length of an angstrom and then think of light as being just a few of these and take 1/80 of that and imagine a surface with hills and mounds no higher than that level! It is pretty smooth!

Zinc selenide is used for a reflecting mirror in some lasers. It has low absorption of the light rays around 10.6 microns. Germanium is also used, but germanium is said to have a "thermal run-away" tendency. It is a relatively high thermal conductivity. The absorption of the energy by germanium does increase with increasing temperatures and so with this type coating it is important to keep the temperature down to below 40 degrees centigrade. An uncoated germanium reflector will reflect about 36% of the impinging energy and that is good.

MORE ON LENSES AND FOCUSING OF LASER BEAMS

It is of primary importance in the use of laser beams that we focus them if we want to extract the energy to do any kind of work that involves temperatures. In the next chapter we explain this in more detail. At this point in this chapter we will examine further some lenses and focusing concepts as applicable to beam-types of light, thus increasing our knowledge of light-control once that light has been generated.

Contrary to some ideas, some laser beams *are not monochromatic* but are composed of many different light frequencies near the very strong center or resonance frequency. This is probably due to some randomness in the stimulation-relaxation action of atoms that may not all move precisely together. Dye lasers may have a broad-frequency beam. Other types may resonate in such a manner in generation and have such filtering that only a very narrow band of frequencies is produced.

When we think of the focusing of the laser beam, we consider lenses that do not discriminate among the various light frequencies as much as a filter-system might. Of course, we know that the propagation of various frequencies through a dielectric medium that is transparent to those frequencies will offer more delay to some frequencies than to others. In this concept we can consider that only certain light frequencies will be focused precisely.

Indulging in some speculation, however, we consider the light from the sun to be nonmonochromatic and therefore composed of many frequencies. Yet we know that we can focus at least some of these frequencies with an ordinary magnifying glass so that we can concentrate the energy in them to such an extent that we can burn holes in wood or paper. In this sense, we would consider the focusing system to be nondependent on light frequency, or perhaps we should say, is adjusted to a given band of light frequencies when it is focused. So we find that light is essentially energy, which travels in electromagnetic waves and has wavelengths and frequency—even though we sometimes consider light photons as bursts of energy or "light-packets" called quanta.

Light waves are said to propagate, like radio, radar, or microwaves, in a direction perpendicular (normal) to the wave front as we have indicated with the Poynting vector earlier. Light travels at different speeds in different materials and the ratio of the velocity of light in a vacuum to that in a given type of material is known as the *index of refraction*.

Some eons ago yours truly had trouble with this concept of a "changing speed of light." It did not make sense to say that at some times the speed of light was slower or faster than at other times.

Everything we know is based on a constant speed of light, some 300,000,000 meters/second. The resolution of the problem was found in a sage's statement: "The speed of light is constant over an infinite distance." Therefore, one can find variations in the speed of light along the way. I tried to imagine a thread that extended to infinity and to imagine a "kink" in that thread. When viewed from an infinite distance the "kink" did not seem to exist at all but it was still there. That "kink" could be interpreted as being the change in velocity durign the infinitesimal part of a second during which time that ray might pass through a different composition medium. Unless you adopt this philosophy, how can you explain radar wave-guides and laser cavities and such?

The basis for lenses is *Snell's law* which says that as light moves through one medium into another is bent according to:

$$N \sin \phi = N' \sin \phi'$$

If you draw a line which is perpendicular (normal) to the side view of a lens, the angle (ϕ) is the angle between the direction the light ray takes and that normal-line. This is called the angle of incidence, or entrance angle into the lens. The outgoing line of the ray is at another angle called phi prime (ϕ') and this is the angle of refraction (bending) between the emitted ray and the normal-line through the lens. (N) is the index of refraction of the material for the incident ray, and N prime (N') is the index of refraction of the material of the lens. Essentially what we mean is that the first N is the index of refraction of light in air approaching the lens, and N' is the index of refraction of the changing medium of the glass lens. Light will bend toward the normal-line in the higher index of refractive materials, which in this case is air. It bends away from the normal-line in the lower of the two refractive materials, in this case the lens.

When an optical engineer beings analyzing the ray progression through various lenses he makes some assumptions. First, he assumes that the angles are small and so the sine value can be replaced by the angle in radians. Second, he assumes that the lenses are thin. Be aware that there is a whole engineering course that considers thin lenses and thick lenses and these two are not quite the same. Also note that if you use some trigonometry with Snell's law you can trace, pretty well, the rays through any planned optical system that uses thin lenses with relative ease.

Now, let us consider a thin lens and an object or light source that is located at some distance (x) to the left of the lens and the light rays from this object or source are incident on the lens through a distance (y). Examine Fig. 7-24. If an object is located off the normal axis of the lens and the object is then reproduced in a larger form some distance beyond the lens, after the light rays reflected from the object pass through the lens as shown in Fig. 7-24, then if the image is larger than the object (as seen by the lens) and we say a magnification has taken place. This can be expressed mathematically as:

$$\text{Magnification} = M = \frac{D'}{D}$$

This equation says that the ratio of focusing distances on each side of the lens govern the magnification "factor." Find D and D' on the diagram and note the difference in distances. You should be able to measure in centimeters the different distances and thus gain some idea of the amount of magnification of the lens.

It is then constructive to think of the situation where the lens might be so that the right-hand focus and object are really a long distance away from the lens. This might be the case where you use a slide-projector. The magnification then is really tremendous. You know that the negative of the film is close to the lens and in front of the light source that causes light rays to pass through the film negative and out through the lens to a distant focal point on a screen. There you'll see the magnified image of the film. You also know that if the negative is "up-right", the image comes out "upside-down". So, it is customary to turn the negative "upside-down" so the image on the screen comes out "right-side-up."

There are situations where more than one lens is used in a focus situation. In this case you simply use the focused image of lens

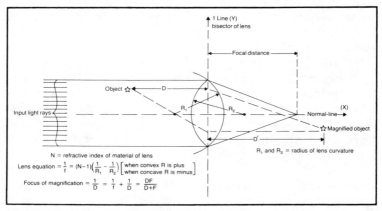

Fig. 7-24. Focus and magnification of an image with a thin lens.

number one to be the input to lens number two and cause the image to be at the proper position to focus into lens number two. There is also an equation for two thin lenses:

$$\frac{1}{f} = \frac{1}{f_1} + \frac{1}{2} \frac{1}{f_2} - 2 \frac{S}{f_1 f_2}$$

where (S) is the separation between lenses. Figure 7-25 illustrates this situation. We have shown the details of curvature extended beyond just the lens outline. A compass is used for this purpose. If you can get the curvature of any lens, open your compass so you can reproduce it properly on graph paper, then you can get a side view of your unknown lens, and using the lens-maker's formula (or as we called it, *the lens equation*) of Fig. 7-24, you can determine the focal distance of that lens. You can probably find the N value at your library if you know what kind of material your lens is made of, glass, silica, quartz, etc.

There is a phenomena we must be concerned with called *spherical aberration*. Quite simply, this means that the focus point as determined by the lens equation and the actual focal point (found from experiment or practice) are not the same. The rays tend to focus at a shorter distance than the equation says they should. This is usually true with thick lenses. The abbreviation for this phenomena is (SA) and it is quite appropriate.

Next, we find that in various types of lenses (some of which are called plano lens types, because they are flat on one side and curved (concave or convex) on the other) that the spherical aberration is different for the different types of lenses. On a plano lens for example, the SA is minimum if the flat side points toward the focal point. This means that you place the lens so the curved side faces the object or light source to be passed through the lens. When using a lens of the type shown in Figs. 7-24 and 7-25 (double convex types) the SA is minimum when the focal distances D and D' are equal. Finally, if you use two plano lenses placed so that the curved sides face each other, you can get a magnification with minimum SA. In this situation you want all the light rays to emerge from the curved side of the first lens in a parallel manner and enter the curved side of the second lens and depart the second lens's *flat side* to converge and thus focus to a point. Of course, the image object that produces the initial light rays into the first lens's flat side must be properly positioned a distance away from that lens so that you have this image or object at the first lens's focal point.

There are some lenses that are not smooth curves, but are non-

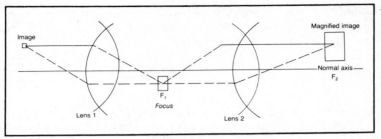

Fig. 7-25. Definition of the focal point of two thin lenses.

smooth curves over their entire face. These consist of curved parts or portions and straight line curves or portions all joined together. The lens is cut this way to minimize its sphericity. They say such a lens is nonspherical in shape. The proper term is *aspheric*. The reason for making such a lens is to reduce the focal distance and minimize the SA.

We know that in camera lens systems there are many lenses of different types, plano, double convex, convex-concave, and double concave. It is not uncommon to find seven or more lenses grouped together in a camera lens mount. That is why they are costly. But the reason (at least one reason) for doing this is to minimize SA and provide good focus at proper distances for the camera body. Each lens must focus into the next in a proper manner. A study of camera lens systems in other texts is recommended if you want to get into the subject further.

The term *chromatic aberration* would naturally have to be concerned with the color or wavelengths of light. Recall that a monochromatic system is a one frequency system or uses one color of light. If we have a lens which tends to separate the focal distances of various colors or wavelengths of light, due to different speeds of travel through the lens for these different components, then we have what is called chromatic aberration. There are lenses that are so designed that the wavelengths of light, over a given band, will all travel the same speed through the lens and so focus at the same outside point. These are called *achromatic* type lenses.

Now we examine a concept of the use of lenses in a system that uses laser light beams. One beam is to excite or "pump-up" a dye laser and the second beam emits from the dye cell laser device. Notice the use of lenses for focusing the output beam to provide a rather wide beam, which then can be filtered to produce a mono-chromatic light ray of the right frequency to excite the dye cell's atoms. See Fig. 7-26. Notice that lens 1 and lens 2 comprise a kind

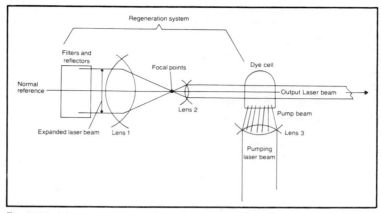

Fig. 7-26. A lens arrangement for a dye laser beam.

of telescope arrangement and they act in a reciprocal (two direction) arrangement. Lens 3 is simply a focusing double-convex type of lens used to focus the input excitation beam into a smaller more concentrated beam to excite the dye-cell atoms.

Of course, there are other types of lenses and different shapes of lenses. The purpose of any lens is to provide a focal point for input and output. We know, and shall learn more in the next chapter about the use of laser beams in commercial applications. Lenses are necessary with laser beams to cause the energy to be concentrated precisely where it should be concentrated for whatever purpose the energy is to be used. **Do not think that because a laser beam is not focused that it may not blind you. It might! Always be careful when experimenting with any kind of focused light or laser beam. Don't look directly into it, —EVER!**

It is of interest for us to examine an arrangement of filters, birefringent crystals, mirrors and polarizers that have been used in a system of laser spectroscopy. We examine Fig. 7-27. We have

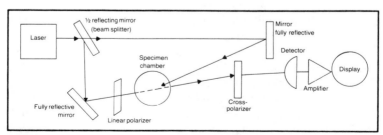

Fig. 7-27. Using light control elements in a laser spectrograph.

previously examined some types of these elements and discussed how they work. We note here that many types may be used to bring about some scientific operation that will, in turn, provide us with information or data or cause certain happenings upon which we can build new hypotheses. It is therefore essential that we not only be able to create laser beams of various frequencies and types, but also to control them by focusing, reflecting, splitting, etc., if we are to gain the advantage of having such beams. Any study of lasers should also include all elements necessary for the control of light.

As we consider lasers, we also think of the range of size now available and the power of the various sizes and types. I am reminded of a small laser that was used in the development of a model train speed control system. It was made by Harry Erwin of the Johnson Space Center in Houston. Using a reflector on the model train's engine, he was able to use the reflected beam of laser light from a gallium-aluminum-arsenide semiconductor laser to determine the speed of the model in feet per second. Such information, being very precise and accurate, enables such scientists to model systems which can be used in space docking of such vehicles as space shuttles and the cargo vehicles that carry equipment and material for space stations. We know that the Russians have a system that provides automatic docking for their spacecraft to their space station that is presently circling the earth.

This tiny laser unit would be ideal for the light source for an optical-fiber system discussed earlier in this book. The GA-As laser beam is in the infrared region of the light spectrum and this type laser is used in modern surveyor's equipment.

Some information contained in the laser beam comes from modulating the beam with five different audible tones. The laser beam itself is used in a ranging system concept, using the light source to "flash" a light pulse to the train and then measure the time it takes for the echo to come back. Doppler information can be extracted from the beam (and tones as well as range) and so both distance and speed can be derived from the returned echo when this is analyzed by suitable reception equipment and associated computers.

In the early experiments of docking with space vehicles, microwaves were used successfully. They still can penetrate some portions of the atmosphere that might be opaque to laser beam rays. Being very small in wavelength (the Ku-band is 12 to 14 gigahertz) the microwave systems are able to determine position and range down to within centimeters of distance—good enough for docking purposes!

We will be very much concerned with the focusing of laser beams in our next chapter. We want to concentrate the beam energy so that we can use a laser beam for cauterizing, cutting, and welding purposes. With this in mind, we examine once more a system which uses a lens for focusing a laser beam. In this experiment, described in Scientific American, May 1979, the laser is used to study some chemical reactions, or cause them. We are interested only in the use of a convex type lens to gather the incoming laser rays and focus them down a point inside of a vessel used as the target container. See Fig. 7-28. In this diagram we consider that the beam (as it emerges from the laser generating unit) is a parallel type of beam with some width, (even though this width is small) and thus may be focused down to a smaller size for the required energy transfer for the experiments. The dotted lines indicate the focal point f_1 to the left of the lens, a condition assumed for the convex type lens, and the second focal point f_2 which is to the right of the convex lens. This is correct if we assume that the light source of the laser is very small and adjusted so it is at f_1. To focus a parallel-beam as just mentioned, might require a plano-convex lens system. The flat side of the lens accepting the parallel and somewhat wide beam, and the rounded convex side then focusing this beam to the point in the container.

The concept of laser weapons has been discussed and analyzed by far more experts and scientists than one can imagine. The idea of a "ray gun" that one sees depicted on TV screens in the "futuristic movies" at times seems within our grasp and at times seems just beyond our capability to fashion. Perhaps this is good, who knows? In any event, when one combined both space technology and the laser development, then the result brings to mind some real possibilities. If laser beams are reduced in energy content by clouds and by atmospheric molecules, then, if one goes into space beyond the atmosphere where there are no clouds or atmospheric molecules to speak of, one has the ability to use a laser easily. This

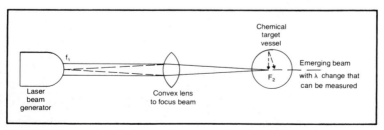

Fig. 7-28. Using a lens to focus a laser beam for chemical experimentation.

is the thinking, past and present, of some in the scientific community. The idea is to mount lasers on satellites and automatically point them at the targets.

Rumor has had it that Russia has made use of "killer satellites" and that these have had some kind of beam weapons as part of their armament. Verification of that might be difficult. To place a gas-laser (one which produces a laser output by abruptly cooling a very hot gas and using mirrors for regenerative reflectors) in orbit may be more costly than they are worth. The equipment required for a powerful laser beam might be more than a satellite could carry. We know that large amounts of power can be present in laser beams because one can stimulate countless atoms to an excited state by using various techniques. These techniques range from electron discharge molecular collisions, the heating-cooling of gasses (flourine) combining such chemicals as hydrogen and flourine, which produce laser excitations upon combination. But this all takes equipment and equipment is heavy and the power required is extremely large. Thus, there is a question as to the ultimate feasibility of laser weapons in space.

But, we can use some of the enormous energy contained in a laser beam to do useful work as we shall see in the next chapter. We can still imagine and wish—and as someone has said "the wish is the birthing of the animal"—that we could make a laser weapon—at least *some* people make that wish!

Chapter 8

Applications of Lasers
and Laser Communications

We have stated that there are possible military applications of lasers. Some of the laser types considered for this use are the "hot-gas" lasers in which a gas is heated to a high temperature, exciting the molecules, and then abruptly cooled. This amounts to a "pumping" situation of the atoms to achieve a laser output. Passing a gas through an electric field can cause the "excitement" among atoms which is necessary to cause them to radiate light photons in large numbers. Once the lasing action begins, it is a relatively simple manner to cause the small initial energy beam to regenerate back and forth in the gas or confined area (by means of mirrors) until the energy gets large enough to pass-out of the cavity or enclosure through a partially reflecting "window." In the military application, the desired use is almost always one of destruction, although ranging and aiming control and position control certainly are usable concepts in the application of laser light beams.

It is of interest to note that the damage resulting from a laser beam hitting a given target is due to the fact that an enormous amount of heat is generated at the point of impact. It is said that damage is caused only by that fraction of energy absorbed by the target. If the target reflects the beam and does this efficiently, the amount of absorption is minimized and the damage capability is vastly reduced. It is stated that the absorption of $1,000$ watts/cm^2 or about $1,000$ joules per cm^2 can melt a metal surface a few millimeters (hundredths of a meter) thick. A pulsed beam of laser light energy might cause evaporation of the metallic surface of a target and the "action equals reaction" concept of the metal flying away from the target might produce a reverse strain and tear a target apart. But, we must consider that to produce such energy on a target would require lasers capable of megawatts of power output.

MANUFACTURING OF LASERS

We are well aware that laser beams have some properties that other light sources do not have. They are nearly or exactly monochromatic and they have spatial and temporal coherence. That simply means that such a beam has continuing "in-phase" properties and polarizations as it travels away from its generating source (assuming no polarization effects from windows and such). It is said that it is the spatial coherence of the laser beam which makes this beam suitable for machining operations. These machining operations can be on a micro-scale, or to think about this in another way, you can machine things down to a fineness heretofore considered almost impossible.

In a modern laser most of the energy contained in the laser beam can be concentrated (focused) to a spot less than 0.001 inch (0.03 millimeter) in diameter when using ordinary focusing lenses. What governs what happens when we focus a laser beam on some kind of material? First we have to consider the power density in the "spot" and second, how long the "spot" is applied to the material. Also, the frequency of the laser light and the kind of material upon which the beam is focused must be considered. What happens is that if we concentrate the energy some material (just like the magnifying glass and the sunlight) the point of concentration will begin to get hot. This may result is just heating the material to causing it to ablate (recall the ablation of the heat shield on the space shuttle?). Ablation is the total removal of the material within the focus area. Thus we can say that the amount of energy concentrated may give rise to an ability to trim some material, scribe on some material, cut some material, or weld some material. In any event, the operation must be held within the focus range of the lens system to get the most concentration of the energy present. In micromachining, lasers are used that have wavelengths of from 10 to 0.5 microns, visualize this as going "down-scale" on a graph from the longer wavelengths to the shorter wavelengths. It is said that wavelengths from infrared to ultraviolet are usable in manufacturing processes.

LASER EQUIPMENT FOR MANUFACTURING PURPOSES

Of course, you need a suitable laser, power supply, and cooling devices. You need to be able to switch the beam off and on at precise moments. You need to be able to position the beam precisely. You will probably need a feedback system to monitor the operation and inform some controller mechanism when the operation is completed. You also need some safety system to prevent damage.

Because certain materials require certain wavelengths of light and certain powers of concentration for various operations, one probably would find that many different lasers might be required in a factory. Lasers have found application in industrial applications in the welding and soldering of electronic parts. They are also used for heat treatment of various machine parts and for drilling holes, which in some applications may be from 0.05 inch to as small as 0.007 inch in diameter. Recall that the laser beam is narrow and can be still further focused so that its "point" of concentrated energy is very small.

One speaks of lasers in the high-power or low-power class and this is like saying that the machinery is "light" for low powers and "heavy" for the higher powered devices. It is considered that the laser dividing line may be on the order of ten watts output. We know that there are many solid-state lasers that produce the smaller outputs, yet they produce sufficient energy to cut through materials such as metal, cloth, and plastics. They may shape ceramics as used in the electronics trades.

Ruby lasers are also used in manufacturing and the following describes how they operate. One end of the ruby rod is a one-way mirror and the other end is a pass-through window-mirror. When the light flash stimulates the atoms in the ruby rod, the light energy flashes back and forth inside the rod until it gains sufficient energy to pass out the "window" end of the rod through the focusing elements. Ruby lasers used in some manufacturing applications operate on about 0.69 micrometer wavelength. There are other solid-state types like the neodymium-doped yttrium aluminum garnet lasers that operate on about 1.06 micrometer wavelength. These wavelengths are in the infrared region.

A ruby or garnet laser can operate with pulsed light flashes to heat a work surface. Other types such as neodymium-doped yttrium-aluminum garnet types might also deliver a steady beam of laser light, or be pulsed if this is desired. This latter type can be pulsed at a fast enough rate so that modulation of the laser beam can be used for communication or data transmission applications.

Gas lasers such as argon and carbon dioxide types operate well in a continuous fashion because they have cooling systems. Some gas systems such as those envisioned for military applications such as the heated-gas (flourine) and electronically charged gas-discharge types are not considered for manufacturing applications in general. A survey has found that many "heavy duty" and high-power lasers are of the carbon dioxide type and can be operated continuously.

The laser beam delivers a highly concentrated amount of energy in a short space of time to a very small area. In many situations this is important, as it can cause less absorption of the heat by the remaining material of the workpiece and thus less damage or "heat effect" on the balance of the work area of the material. There may be no cracks or metallic distortion or the setting up of heat stresses in materials, which sometimes happens when the material is subjected to broad applications of heat to the surfaces. Also, due to the high concentration of energy, the material to be removed from a "cut" or a hole may be vaporized and thus there are no droplets of material to be removed.

In some applications a laser beam is so connected with a source of oxygen that when the heat is generated by the beam and the oxygen is applied, the result is a burning of the material and this can produce fast cutting or shaping of the material. Suppose that you have a material to be deposited another alloy of some kind. Using a laser beam, properly adjusted, and feeding the alloy in the proper manner, the beam of the laser can heat it to the melt point and deposit it upon the receiving surface. As the laser energy is well concentrated, not much heat is delivered to any other area and so the operation is really brought down to a "point source" type of application.

There are, of course, innumerable applications of laser cauterization in medicine, and this is a step forward as one doesn't heat up other tissue nearby. "Tacking" tissues together as in eye operations are now quite common, and danger is reduced because of the fine plane-of-focus necessary to concentrate the light energy. Daily we learn of new applications in almost all fields of the use of laser beams.

SOME DEVELOPMENTS IN LASER TECHNOLOGY

In the field of medicine spectacular developments seem to be underway using micro-size lasers. Some researchers have used tiny lasers attached to catheters to remove fat obstructions in the arteries of animals. The marriage of camera-like fiberoptics to small optical fibers which are inserted into catheters enable medical researchers to probe deep inside the human body and examine many areas heretofore visible only after extensive surgery.

We know of the use of lasers to open tissue and seal it and thus with the micro-miniature laser types now available, and the use of penetration devices such as catheters, it would pressage a time when internal surgery might be accomplished without opening the body. Reduction of infection, smaller incisions, etc., thus seem the "spin-off" of the use of laser technology in medicine.

One method of tuning a laser has been patented and this concerns the use of a piezoelectric crystal. The patent describes how the crystal, which has a reflective surface that can be deformed with electro-acoustic waves, can be so changed that it acts like a diffraction grating and thus can filter beam wavelengths so that various ones might be amplified.

Switches for laser beams have been made from acoustic-optical cells and their associated optical components, so that a fast "chirp" (changing frequency wave) in the cell will cause the light beam to be diffracted so it can be matched by the transfer function of the optical system. When the diffraction is not precise the optical "match" does not take place and so a "Q" switch or very fast optical-shutter effect takes place. It is of interest that a high-power "Q" switch has been developed. This switch has been used in a 100 megawatt ruby laser cavity. A Frustrated Total Internal Reflector (FTIR) is pulse driven by a mechanical shock-wave to produce the fast switching action.

The "steering" or positioning of a laser beam is desirable in many applications. You want to change its direction as a function of some other quantity so that it will perform some peculiar or particular task. The question, then, is: "How do you "steer" a laser beam?" The Hughes Research Labs of California has come up with a method. They have developed an electronic beam scanner for a CO_2 laser to meet the following scientific requirements. They wanted a 3 degree scanning range; 100 resolvable beam positions (think how small a measurement that would have to be for a maximum scan of 3 degrees). They wanted a 6,000 Hz raster scan in one dimension, and a capability for 100 watts of 10.6 micrometer wavelength output.

In the device developed, they employed an optic Bragg diffraction grating using germanium, with a acoustic-longitudinal acoustic wave, and a laser beam polarization in the crystal (111) direction. The numbers refer to the crystal's axes. A sawtooth wave of the FM type over a 27 MHz band centered near 100 MHz—a 27 MHz deviation—was used to provide the scan signal and mechanism. A small transducer provided sufficient beam divergence in an acoustic mode to produce the 3 degree scan capability. Direct contact water cooling of the Ge crystal was provided to prevent thermal run-away. Cylindrical, reflective, telescopic optics were used for beam conditioning and shaping recollimation. The transducer bonding was done ultrasonically using thin metal films of Ag, Au, and *Ag-In*. The latter gave the best bonding performance.

In a program of study done by Hughes Research Labs at Malibu, California, an attempt was made to determine the best possible modulation system for a CO_2 laser. This laser operated on 10.6 micrometers and was to be intensity modulated. That brings to mind immediately some kind of shutter or opaque-transparent arrangement or device that can be varied in ability to pass the laser beam as a function of some modulation current or voltage.

The Hughes performance requirements were a minimum frequency response of 200 MHz and a modulation capability of 100% with a minimum of driver power. Thus they began a serious review of all known modulation techniques for laser beams. We find this listing of value to us in the futherance of our knowledge of laser modulation techniques. Some of the methods considered were the electro-optical effect, Stark Effect, acoustic-optical effect, and free-carrier related effects. We note that the electro-optical and the acoustic-optical effects would imply that the passage of the laser beam would be affected by either an electric current, which somehow would affect an optical device, and an acoustic wave, which in turn was probably produced by an electric current varying at some intelligent rate and in some intelligent manner. Of course, there were other techniques that they explored that either weren't successful or were of such a nature that divulgence of the process was not in their best interests.

Through an elimination process the techniques that did not have the 200 MHz modulation capability bandwidth and those that required too much power to make them effective were eliminated. This left some techniques that all embodied some kind of optical effect as controlled by electronics. These turned out to be an optical waveguide, a microwave bandpass traveling wave, and intercavity coupling methods that use microwave techniques. The changes may be induced electrostatic or electromagnetic fields or physical changes in capacitance or inductance of the cavity or merely its resonant size. Also, a baseband traveling-wave (TEM) parallel strip and multiple pass lumped elements were used. It was found that the intra-cavity coupling was very attractive as a flat bandpass in excess of 500 MHz was produced.

CONSIDERATION OF BRAGG'S LAW

This law predicts the conditions under which diffracted X-rays from a crystal are possible. As it turns out, if we have a crystal (such as calcite) that can refract X-rays it can be considered to have any number of refracting planes within its structure. The X-rays are considered to penetrate the crystal and to be refracted from these planes. The "trick" is to so refract them that they add, constructively and thus form a beam. Bragg was able to derive an equation that proved that the beam formation *could* occur if the spacing between the planes in the crystal were an integral number of wavelengths, i.e., (1), (2), (3) etc., and his equation (which reduces to a simple form) is:

$$2d \sin\theta = m\lambda$$

where (d) is related to the cell dimensions by:

$$d = \frac{a_o}{\sqrt{5}}$$

If, then, X-rays fall on a set of atomic planes so that the above equations are satisfied, that is, if:

$$\lambda = \frac{2d \sin\theta}{m} \text{ (m is an integer)}$$

and λ is the X-ray wavelength, then diffracted beams of X-rays will result. Sodium chloride, for example will produce a refracted beam at (θ) equal to 12.6 degrees for (m=1).

OPTICAL SYSTEMS DATA TRANSMISSIONS

Assuming that you can modulate a laser beam satisfactorily with the required flat band-pass and depth of modulation required to convey signal data or intelligence, then what can we call this process? We know that when we combine two signals we can have hetrodyning. Thus we are not surprised when we learn that experimentation has gone forth to combine in multiple hetrodyning fashion microwave signals with infrared or optical signals. Such a data modulation system uses the multiple passes of the infrared or optical signals through a crystalline material that also has an acoustic-optical signal applied in such a way that it makes the crystalline material become an acoustic-optic grating. Meantime, in the growing of the crystal and preparation thereof, fixed gratings are fashioned so that the optical signal or infrared signal is properly guided through the material. The output of the device is composed of the input signal, infrared or optical, that has become modulated (varies in intensity) with the desired data to be transmitted. This signal can be transmitted optically, and here, optical fibers or optical-fiber cables may play an important role. The optical signal can be converted into an rf signal for transmission.

LIGHT BEAM SCANNERS

If we have a laser beam and cause it to impact with a piezoelectric crystal it can be made to perform as a traveling-phase grating. This will diffract the monochromatic light beam. This traveling-phase effect is made possible by an acoustic-wave that propagates across the crystal to produce the grating phenomena. The light beam from the laser is diffracted in accordance with the acoustic wave deformation of the crystal. The light intensity at some first order spot is detected to provide a temporal signal representative of the light distribution of an image line. If the image is in colors, then by using a *chromatic* beam, one may separate the three primary wavelengths to re-create the image in the scanning beam.

A RECONSIDERATION OF OPTICAL GRATINGS

It is time to reconsider this important light-control element. We find it used so very much with laser beams, and we find it in so many diverse forms that a review of the "grating" is beneficial. When we have two slits, as in Young's experiment, we assume that such slits are small compared to the wavelength of light passing through them, and so the light is diffracted rather uniformly. This means "bent" away from its original direction of propagation. We examine Fig. 8-1. We are not surprised to find that the screen upon which the "slit-beams" fall has light and dark spots upon it. When the radiations in the beams are "in-phase" a reinforcement takes place and we have brightness. When we have an out-of-phase condition we have darkness of the screen.

So, it turns out that we can have a wider system of diffraction (bending) of the rays of light if we use a spot source and a lens and a wider slit. If no lenses are used this is called the Fresnel diffraction. If lenses are used before and after the slit it is called a Fraunhofer diffraction. So we come to the *diffraction grating* concept. If we have a large number of "slits," (or as they are more commonly called "radiators") then we have what is called a diffraction grating. In this case the larger number of slits enables the propagated rays to mix "in-phase" or "out-of-phase", or at some phase relationship be-

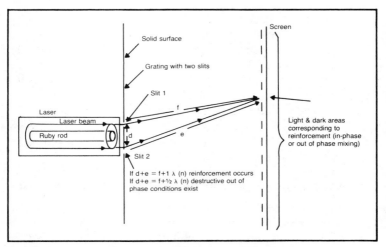

Fig. 8-1. A two-slit defractor as per Young's experiment.

tween these extremes. The result will be light and dark lines or spots on a screen (if one is used). The diffraction grating is a device that performs the equivalent operation of a multi-slit window for light rays.

Thus we can imagine how using some kind of element that provides this "slit" effect, will control the amount of light passing through it. Diffraction gratings can be made from glass that has been lined or scribed with fine lines of a proper depth and thickness so they act like slits. They can also be made of films or various other substances that can reflect or transmit light rays through them.

Not far from where we live there is a traffic signal that has a diffraction grating of a rather large size covering each light. As we approach the signal we cannot determine which light is on and which one is off. But, when we get close enough and in position near the intersection of the two roads, we suddenly find we can see red or green quite clearly. This is a diffraction grating system in front of the lights. It passes the color accurately, and passes them *only* to that distance wherein the grating-plus-lens is focused.

In any event, a diffraction grating is very important in light control. The ability to vary the grating so it will pass more or less light as a function of some electric current or acoustic wave or such other phenomena as heat, pressure and so on, makes it possible to develop signal systems, alarm and sensory systems and convey data by amplitude modulation of the light beam generated from some laser source on LED.

Gratings are made by ruling equally spaced parallel lines and have depth. If a metal plate system is used it is called a *reflection grating* as light does not pass through it but is reflected with all the interference effects of the "pass-through" gratings. To cut the grooves precisely it is usually necessary to use a machine with a diamond cutting element. Once a grating has been made from some metal, then plastic films of various types may be coated on the face, stripped off and the film used as the grating itself. Also, we know that a piezoelectric crystal might be used as a diffraction grating for light control in a modulation application.

AN ADVANTAGE OF LASER COMMUNICATION SYSTEMS IS SECURITY

We know that the laser beam is narrow in diameter and thus is not readily exposed to non-authorized and illegal revelations of its data or information. Among the military such a system as a secure communications system using a laser beam as the "carrier" would be ideal in many areas; tank-to-tank, sub-to-sub, airplane-to-airplane, airplane-to-sub and so on. One problem is to get that laser beam to propagate through air and water over desired distances at sufficient energy levels so that the basic reason for doing so can be realized.

While it is difficult to propagate laser beams through the atmosphere with any amount of destructive power—at least at present—they can be propagated with sufficient energy so that they can be detected and can transmit or carry information and data. One proof of this capability is the ranging systems using laser pulses now in existence and the surveying instruments using laser light. For shorter communications systems internal to vehicles in hostile environments of any type, the use of optical fibers promises a secure system as we have mentioned in the earlier part of this book.

It becomes of interest to realize that there is one phenomena of the laser beam which also helps to prevent security leaks in communications and data systems. That is the polarization effect. For various reasons the polarization of a monochromatic beam of light may be fixed at some definite, relative position. It may then have to be rotated through some required angle in order to be receivable in proper form. Thus it may be possible that interception of a laser beam (that may have some peculiar way of handling the polarization of its constituents as a function of time, message, or type of data) may be somewhat difficult.

In this same sense, another concept relating to beam security is the tunable dye-laser that has an acoustic-optic beam deflector, a folded *Lyot* filter and a diffraction grating that altogether produce a discrete, periodic, or digital type output. The periods between the discrete wavelengths in the output of the laser are varied electronically with the input data or intelligence to be communicated. Recall that the dye-lasers can be operated continuously and thus provide a good basic means of continuous-wave communications.

In the quest for fast "tuning" of lasers, which means adjusting them for close-but-different wavelengths of output, an organic dye laser may be used. The tuning element may be a diffraction grating and a Bragg diffraction cell driven by an rf source, which modulates an ultrasonic transducer. Notice here the use of two elements we have previously discussed as important in light-control systems.

SAMPLING A LASER BEAM WITH ACOUSTIC PULSES

There is a patented method for sampling and scanning a laser beam with acoustic pulses. The pulse is propagated across the light beam and the *pressure variations* (of air) which comprise the pulse tend to form a diffraction grating for deflecting the light beam over its volume of interaction with the pulse pressure waves. As the pulse is propagated, it doesn't all propagate instantaneously and so, due to the gradual movement of the pressure wave through the beam, the diffraction effect takes place gradually and so there is a scanning effect produced. If then the beam is focused through suitable optics and on a suitable detector, an analysis of the beam in its parts may be undertaken. This may be useful in analyses of various chemicals and elements.

A VIDEO-LASER RECORDER

Ampex is and has been working on a video-laser recorder that will be a two channel 16 mm laser beam film recorder for television type sensors or cameras. The two channels of video at 525 or 875 lines per frame are to be recorded at 30 frames per second with the frames on side-by-side format on 16 mm film. The system is said to have a neon-helium type of laser with its necessary optical system. A 70% efficiency was recorded at 16 MHz bandwidth. Here, one would imagine a TV camera modulating a laser beam, which is coupling its light output to a film record system. You can imagine the amount of equipment necessary in such a system, but it is another use for a laser-type light beam.

THE LASER BEAM PHENOMENA IN MODULATION

What seems to be a trend in laser technology and one that

seems to "boggle the mind" is the fact that acoustic waves can cause an effect which causes a laser beam to change direction (diffraction) or can cause modulation of the light beam. Somehow we don't always think of the combination of acoustic or pressure waves with rf or electromagnetic energy that seems to occur in these instances.

One must realize that within our atmosphere, the air molecules that normally absorb and impede the radiation of light rays or other rays in the "higher" electromagnetic spectrum, can be influenced by various devices that may cause them to contract and expand and so on. Therefore, it is not beyond reason to say that if the air is caused to change, due to acoustic or pressure waves or due to some transducer device, that light, then passing through this air, shall be affected. The absorption rate would change, the mixing of the pressure wave front with the electromagnetic wavefront—a complex mathematical equation—suggests that the beam of rf waves will be refracted or deflected as the case may be. Furthering this concept, if both the acoustic wave and the light wave are impressed on some "transfer device" such as a crystal, mixing takes place that causes interference patterns. If we then shift the beam from a detector in an on and off manner or vary its intensity as a function of some variation in the acoustic wave, we say we have "modulated" the light beam.

THE LASER IS A RATE DETECTOR OF ROTATIONAL SYSTEMS

Strangely enough it is possible to add to the velocity of propagation of a laser beam the very tiny velocity of roll which may be due to the turning of a rocketship, airplane, satellite or whatever. Then we can extract information to enable us to determine precisely what the roll-rate of the object happens to be. The sensor for such a system of rate detection is a *laser interferometer* and we detect the fringe shifts caused by the body's rotation. Here is a somewhat technical explanation of how the system works. A light emitting-diode pumps an optical-fiber laser (a laser bead at one end of an optical fiber). This sends light rays through the optical fiber or distribution waveguide. The coherent radiation rays from the laser are then partly backward coupled by means of an acoustic-optical grating into a reference optical-waveguide (optical-fiber). The coherent radiation is also coupled into each end of a multi-turn optical waveguide positioned so that it serves as an interferometer. A pair of directional couplers, one for clockwise radiations and one for counterclockwise radiations will extract the coherent radiations from the interferometer fibers after the light rays have made at least one passage around and through the device. Finally, having both the clockwise and counterclockwise signals, these are combined and the phase shift between the two signals is measured. It turns out that the amount of phase shift is directly proportional to the rotational rate of the system holding the interferometer.

This seems like a hard way to accomplish the rate determination but we need to think of the advantages. First, a gyroscope, which also can measure rate by the rate of its precession has a tendency to drift so that its axes do not align perfectly with the axes of rotation. Thus an error can be introduced. Second, a gyro wheel will have friction. Any variation in its rotational rate will affect the precession rate and thus the accuracy of measurement.

We recall, many years ago, when we examined a rate gyroscope designed by the German scientists at Peenemunde, to be used to cut-off fuel flow in the V-2 rocket system. If a high degree of accuracy was to be obtained, it was necessary to constantly determine the velocity of the rocket and shut down the fuel at a precise rocket velocity so that a precise range would result. It was fortunate that such a device did not appear, at least in any numbers, in the rockets launched against London. Instead of the ten-mile errors of the first V-2 rockets, improved types could have hit, directly "on." As we view this laser-type interferometer, again the accuracy of such weapons of war come to mind and the guidance of them. Precise accuracy over millions of miles of space is now possible.

THE YAG LASER

This is, of course, the yttrium-aluminum-garnet type of solid-state laser. We find it mentioned by the abbreviation (YAG) in much of the current literature. Some YAG lasers operate at about 1.06 microns in the near infrared portion of the spectrum and they may be used in ranging systems quite effectively.

A LASER FLAW-DETECTION SYSTEM

The method used with lasers and opto-acoustics replaces a type of flaw detection system used in industry that uses piezoelectric transducers. A flaw in a system is detected by the transducers that are physically in contact with the surface to be examined. Acoustic waves change phase according to the flaw and thus may be detected.

If one uses a pair of lasers and a suitable optical system, it is possible to use one laser to generate a high-powered pulse that is converted by the material into an acoustic wave in the material. The second laser is used as an interferometer that will monitor the surface motion and thus the detection of the flaw is apparent by the phase shift as previously described. This system needs no direct contact with the material to be examined, nor is there any limitation on the size or shape of the object under test. The speed of examination is very fast, according to the Harry Diamond Labs at Adelphi, MD.

What seems to be almost impossible, yet is very possible is that one might communicate from a satellite to an underwater submarine through the use of underwater acoustics and a scanning laser beam. It seems that the pressure waves from the submarine's acoustic transducers can cause the ripples in the water that normally are so small and fast they are undetectable. Due to the infintesimally small wavelengths of laser beams and the fact that they may be propagated with a given polarization, such physical disturbances can be detected either through a change in polarization, or phase differences or similar phenomena. This same idea of tiny measurements with a laser beam are also much in evidence in a study of geophysics wherein minute changes in the earth can be determined, magnified and studied. The laser beams can represent the sensing element for such studies.

Lasers also help to align sawmill blades, guide the cutting of marble slabs, and even position patients for various medical procedures. In addition the use of laser beams from microscopic size laser generators can, according to recent reports, clean-out arteries and veins of humans by removing fatty deposits or breaking them loose. They also can cauterize human tissue such as the retina in the eye-ball, cut away undesired tissue such as cataracts in the eyes, and furnish light sources for inspection inside the human body. We have already mentioned "drilling" holes, and welding, soldering, and cutting usages in industry.

LASERS IN COMMUNICATIONS SYSTEMS

The use of light beams for communications is a development that finds great favor among many scientists and researchers. The fact that the equivalent of a "hard wire" direct coupled system might be made that has no wires, immediately brings to mind security possibilities such as in airplanes, battleships, submarines, on battlefields, or over distances that are so great that other types of communications cannot be easily installed, handled, or maintained.

We think of a laser beam as being a radiation in the electromagnetic spectrum. Then we can think of the use of modern modulation techniques to convey intelligence over such a link and we can conceive of adaptations of "normal" communications equipment to facilitate this communication. A case in point is the development of a 10.6 micron infrared superhetrodyne receiver front end for use in a wide-band CO_2 laser communications link. The infrared receiver used an 850 MHz response PV (HgCdTe) photomixer mounted in a suitable housing. An i-f amplifier and preamplifier was also used that had a response from 5 to 1500 MHz. The receiver was designed to accommodate a doppler shift of some plus or minus 750 MHz while providing an instanteneous information bandwidth of some 400 MHz. This unit was tested by the Airborne Instrumentation Labs at Mellville NY. This demonstrates the use of the somewhat conventional types of ideas in fabricating an infrared communications receiver front end.

THE LASER IN A FIBER-OPTIC STRAND

Optical-fiber lasers use distributed feedback to make them operate, and their use is envisioned for communications systems, which employ optics as the basis for the operation. We note here that at the National Aeronautics and Space Administration of Pasedena, Calif. a laser was made of an integral part of the optical fiber channel either by diffusing active materials into the optical fiber itself or surrounding the optical fiber with the active material. The oscillation of photons within the active medium to produce the lasing action (the pumping) was done by making a grating of the optical fiber in such a way that distributed feedback—positive reinforcing feedback, took place.

What interests us is that this is evidence of the joining of the fields of optical-fiber technology and laser technology to produce a device that embodies something of both technologies. We have mentioned using a laser of small dimensions to "feed" an optical fiber, but here we find that the optical fiber itself may be adapted to being a laser and from this start we can easily imagine joining of more optical fibers to extend the range and use of the light thus generated. We easily imagine the possibility of modulating the laser beam with any pertinent kind of information and the reception and decoding of such information at the user end into usable form.

SOME PROBLEMS WITH LONG DISTANCE LASER COMMUNICATIONS

If one is trying to establish a communications system using a laser beam over a vast distance such as might exist between a satellite and a ground station, one of the first problems to be encountered is that of "acquiring" the target and keeping it in view so that the very narrow laser beam won't drift away from it. This may require rather complex instrumentation at each end of the system (for example, a radar to acquire the remote station initially). This radar might be a microwave type. Then a laser type of radar could be used to refine the pointing of the ground communications transmitting-generating system. Recall that you don't want to just illuminate a satellite with a communications laser beam, you want to illuminate directly and accurately a receiving sensor on that satellite. Consider a laser beam to be, say ¼ inch in diameter and the satellite to be some 9 to 12 meters in size facing the radar, then you intuitively realize that a precise focus of the laser beam to one spot on that satellite's surface will be required.

Some scanning probably has to be accomplished to acquire the target sensor or receiving sensor. Some kind of "lock-on" has to be employed through the various equipments at both ends of the laser beam channel, and then the beam can be modulated and demodulated with the desired intelligence.

Some years ago a series of optical communications experiments between a target at 60,000 feet and a ground based station were conducted by NASA. The basic system was an optical tracker and a transmitter that were located at each end of the communications system's link (one airborne and one ground based). The aircraft transmitter consisted of a 5 mW (NeHe) laser with a 30 megabit modulator. The ground station was an argon laser operating on a 488 nm wavelength. The problems of atmospheric degeneration of the system, acquisition, tracking, and such, were carefully studied and many experiments on data transmission were conducted.

Both speech and music were transmitted in excellent fashion over a few kilometers. But this experiment implies a successful modulation system, which, with music of high fidelity must have a good bandpass, perhaps as high as 15 kHz. Speech sibilants require larger bandpasses if they are accurately and completely transmitted. Also one must consider that the recovery system, the receiver, must be capable of demodulating such a modulation system with just as much precision. From this level of transmission (speech and music to video and pulse-data transmissions) much work is still required for the level of perfection desired.

In this connection a system known as Pulse Quaternary Modulation (PQM) has been successfully demonstrated in laboratory experiments. The significant part of this modulation system was that an operational 400 Mbps bandbase laser system might be developed on this concept (Mbps is Megabits/per/second).

In a signal patent for laser communications, a device is described that can split the output beam from a laser into two quadrature polarized beams and then phase modulate one of these beams and recombine the beams in a noninterfering manner for transmission along a single path to the modulator. At that point the beams are received by a receiver and separated and then combining the beams destructively causes amplitude modulation of the combined beam. At the target, a receiver can remove the modulation information from the laser beam and send it to any kind of desired read-out unit. This is another example of how you can modulate (AM) a laser beam.

Back to the problem of laser communications over any long distance. Due to the narrow beam divergence of a laser or other optical communications system, very small angular motions of either end of the system can cause misalignments which could mean loss of data or intelligence. One location or situation in which this factor has to be considered is in the military applications of laser beams for communications channels. In a moving tank there is a lot of bouncing around. In other vehicles such as helicopters and airplanes and such, the wind forces on the body tend to cause minute rotations which normally are so small as to be undetectable by people, but which are enormous deviations of a laser transmitter generator mounted in such a vehicle. As an experiment, try focusing a small-beam flashlight on some object a long way away and then try to run and keep the focus correctly aligned! Highly sensitive and completely automatic tracking systems with laser communications equipment is a necessity. This is in addition to automatic frequency tracking and phase or polarization tracking and generation and power output stability.

One can consider that two othorgonal polarization states of a light beam carrier can correspond to our well known digital states of 1 and 0. In such a system automatic polarization compensation is provided by applying a dither modulating voltage to a cell which exhibits electro-optic effects. A dither is a voltage which varies rapidly and periodically so as to keep the object it controls in motion to eliminate the stiction or viscous friction delays that might occur if one had to stop and start the object each time some intelligence or control signal was present.

The electro-optic cell controls the relative phase of the electric field (polarization) of the input light beam enabling the dither frequency component of the difference of the instantaneous powers in the two polarized signals to be coherently detected. A signal from the detector is fed back to the cell after integration to form a polarization-bias-compensating servo-loop.

A MEDICAL COMMUNICATIONS EXPERIMENT

Some time ago an experiment was carried out to determine if it was possible to convey sufficient information via a laser beam so that a nurse anesthetist located in the V.A. Hospital at Cleveland could obtain instructions and guidance from an anesthesiologist at Case Western Reserve University located some 1.2 kilometers away. In order to be able to handle this medical situation properly it is necessary that television signals, voice signals, and data signals on the physiological condition of the patient, such as electrocardiogram, pulse, respiration rate and such, be accurately transmitted. This was accomplished in the experiment.

When viewing this medical situation the implications are far beyond just the ability to transmit a lot of data over a laser link such as was used in this case. One realizes that nurses as anesthetists administer a large portion of the anesthesia in medical situations. The ability to be able to provide guidance and help via such data transmission systems that are immune to interference may go far beyond the case of just anesthesia. Visual, voice, and data transmission over the same link simultaneously provides a proper input of accurate diagnosis and medical application.

EARLY PUMPING EXPERIMENTS FOR A NdYAG LASER

An experimental program was conducted to determine and evaluate the alkali lamps for use in a space qualified NdYAG laser system. The alkali lamps had bores from 2 mm to 6 mm. Measurements were made of the lamps performance both in and out of the laser cavity. One significant observation was that for a constant vapor device that spectral and fluorescent output did not vary for vacuum or argon environments and this led to the use of this laser type in an argon environment. The alkali lamp was finally optimized at a 4 mm bore and for a 430 watt input the laser produced slightly over 3 watts of coherent beam output.

ATMOSPHERIC ATTENUATION OF A LASER BEAM

The attenuation of a laser beam propagated in the atmosphere has been measured. The laser used was a 10.6 micron carbon dioxide (CO_2) laser and the beam strength was measured by a pyroelectric detector after it had traveled one kilometer through the atmosphere. Using a 20 mW power output level, the focus of the target placed at the focal point of a 25 centimeter focal length spherical mirror is one square millimeter. Rain and haze produced no noticeable attenuation but heavy fog gave attenuations of 3,000 times what would be normal without any attenuating materials.

TIME DELAYS IN OPTICAL TRANSMISSION SYSTEMS

We normally do not consider this to be much of a problem. The speed of electricity and light propagation and the propagation of microwaves is so fast we usually do not worry about it. But there are some applications in which time delays can be very important. Also the transmission of pulses (that contain information either singly or in groups, such as a pulse-position modulation system) may be affected by time delay in the system.

Time delays may occur in the equipment used to code and decode such information, in the actual transmission times of the rays of light, and in the response times of detectors and generating times of the laser units. This can be especially true when we have any kind of off-on program operating the devices. There is also the problem of synchronization required in the correct operation of some electronic-optical systems. Synchronization might be affected adversely if the timing is not precisely correct.

MORE ATMOSPHERIC TESTS

Currently our primary type of data transmission is digital. This fits into our computer patterns neatly and so it is desired. Also information theory suggest that this may be the best way to get desired information from one point to another with the least loss of data and degradation of data. Thus some digital data transmission tests were conducted using a laser beam propagating through the free atmosphere to determine the error frequency and magnitudes. The error frequency was determined using a pulse-code modulator that modulated an argon laser beam propagating with a one watt output and operating on its *green line* of 514 nm (nanometers). The path was short (3 kilometers) and led over build-up areas. The results were not encouraging. It was shown that this form of digital data transmission is suitable only for specific applications.

SPACE EXPERIMENTS WITH LASER
BEAMS AS COMMUNICATIONS-DATA CARRIERS

Having explored the attenuation of laser beams by our own

atmosphere and being aware that we are going into space, the next idea is: How can we use lasers in space applications? Is it immediately apparent that because there is no atmospheric attenuation in space it should be easy to propagate the laser beams there. So communication to and from satellites is a possibility.

The Aerojet General Corporation conducted some experiments along this line and called then the (LCE) laser communications experiments. In the experiments CO_2 lasers were used both in the vehicle and on the ground to establish communication between a synchronous orbiting satellite (meaning it had to be some 23,000 miles into space above the earth) and a ground station. The problems of beam generation and acquisition by unmanned satellites were also part of this study.

If we consider the space environment for lasers and think about the power situation and realize that the sun is ever present in space and that we can convert sunlight into electricity, then we realize we have abundant power available for electronic uses once it is converted from photons to "holes and electrons." If the balance of the electronics devices, integrated circuits and such, can be devised in such a manner that they can withstand the rigors of space, then we have an opportunity to put something into space and use it for an indefinite time. How fascinating!

Of course, you realize that we have had Voyagers whose missions run into years and we have reactivated satellites which did not fail, but were purposefully "shut-down" by their masters. So the concept in a sense has been proven. Now we think, also, of the possibility of ejection of heavy ions from the propulsion system of some kind of space ship and we may have a system of operation and propulsion which can go anywhere and for any length of time, so long as there is sunlight to make possible the creation or conversion of electricity from sunlight. It also makes us take a new look at the technology that is developing at such a rapid rate. How nice it would be for that same technology to develop sunlight conversion units, which unlike our present rather inefficient cells, could operate with an efficiency of 80 to 95 percent. Not only would that technology be acclaimed for the applications to space applications, but also the more mundane applications right here at home.

MODULATING A MICROWAVE CARRIER WITH A LASER BEAM

If one has a CO_2 type laser one might be able to combine its output with that of a microwave system in such a manner that the resultant sideband will have the results of the sun and differences of the two frequencies, and this might be usable in optical image radars and high data-rate systems according to a study by United Aircraft Labs of East Hartford.

The goal of their program was the demonstration of efficient sideband generation of a 10.6 micrometer CO_2 laser carrier at X-band microwave frequencies in a thin film nonlinear GaAs optical modulator element that is interfaced with a microwave ridgeguide structure. The electro-optical interactions in a GaAs thin film between an optical guided wave and either a traveling wave or a synchronous microwave of the standing type was conceived to be capable of producing the sum and difference frequencies of the two interfacing signals. Then, a "chirped" (changing frequency) optical signal is generated by using a frequency modulated microwave signal. This signal can be separated from the others by a proper optical filtering technique such as a diffraction grating or a narrow band-pass interferometer.

SOME CONCEPTS OF AN OPTICAL RADAR SYSTEM

The Jet Propulsion Labs of CIT have shown that an optical radar system is feasible. The system is composed of an optical cavity with a laser and mode-locking means that are used to build up an "optical pulse," which is radiation in the optical spectrum.

An optical switch was provided in the cavity to convert the polarization of the optical pulse generated within the cavity. The optical switch was made of an electro-optical crystal driven by a time-delayed driver circuit, which is triggered by a coincident signal made from the optical pulse signal and a gating pulse signal. The converted optical pulse signal strikes a polarization sensitive prism and then it is deflected out of the cavity toward the target. It consists of a pulse made up of the optical energy generated by the laser during the pulse build-up period. Then, as in all radar systems, after this pulse strikes the target, some of its energy is reflected back to the starting point and through proper receiving equipment its energy is converted into a kind of signal that can be compared to a timing pulse in a timing circuit to determine the total travel time to the target and back. Thus, by proper interpretation of this time, the range to the target can be determined.

It is interesting that using this kind of radar system, the angular measurements may be much more precise than with conventional types of radars because of the narrow beam transmitted. Microwave type radars have a relatively broad beam and thus leave room for a certain region of target-location uncertainty. This error uncertainly is vastly reduced with an optical radar although acquisition of the target may be much more difficult because of the narrow optical beam generated by a laser.

This brings up the ideas of some experimentation conducted by some agencies to try to widen this optical beam for target acquisition purposes. Some have believed that defocusing the optical beam may tend to widen it in all cross-sections, thus making it possible to acquire a target easier than by automatic focusing of the beam.

If this defocusing system is not feasible or practical then some believe that a microwave radar might always have to be used in conjunction with a thin-beam laser as the acquisition device. This device then "points" and positions the laser equipment so that it can then track the target. One then has acquisition of targets accomplished and precision tracking, which is growing more mandatory all the time. Be aware that this precision tracking is not just for a motionless type of target, but a fast moving and even maneuvering type of target—even if the movements are slow compared to the propagation speeds of an optical light beam. We have seen the acronym OPDAR used to mean Optical Detection and Ranging systems. Notice the similarity with the word acroyonym RADAR (Radio Detection and Ranging).

THE CONCEPT OF A MULTIPLEX DIGITAL LASER GENERATOR

An interesting concept is multiplexing. This means imposing several channels of information on one carrier in some manner or another. In this case it is a multiplexing of digital information on a laser beam and its general method of operation is fascinating. Recall that "time multiplexing" which is commonly in use on hard wire computer lines and on radio systems, means that one devotes some space of time to each channel of information *sequentially*. All channels are *not* transmitted simultaneously. Only in frequency multiplexing as applied to audio and rf signals can the combined output of channels be used to modulate a carrier and be transmitted all at the same time and then decoded by appropriate filters, etc., on the receiving end.

In this laser concept of multiplexing, the frequency components of the laser emission are spatially separated (beams are separated) inside the generator cavity. You know how this can be done with a proper prism system. Then each component of frequency, or beam, is separately pulse-code modulated according to a digital input signal, transmitted to the receiver in a recombination of the beams and then separated for decoding. The receiver system is relatively simple. Of course, if the beams were not recombined and were transmitted as separate beam elements or constituents then simple photo-to-electric detector systems or infrared-to-electrical signal detectors might be used for the receiver.

It is well to recall at this time that in a gas laser (CO_2 type) what happens is that the gas is caused to pass over some electrodes mounted transversely to the gas flow. Then a spark gap provides added pumping energy when it is discharged into the gas flow causing lasing action. It is said that the gain-bandwidth of an atmospheric CO_2 type of laser discharge is sufficiently wide (around 4 gigahertz) to permit as many as 20 axial modes to oscillate simultaneously within the cavity. Proper filtering will separate a desired mode of operation.

USE OF THE CO_2 LASER IN PRODUCTION

Making printed circuit boards of smaller and smaller sizes is simplified by using a CO_2 laser to "drill" the appropriate holes for mounting components. The Coherent Everlase CO_2 lasers are used in this operation. We find that this operation for a modern circuit board requires many thousands of holes which are very small in diameter (0.004 inch) to fit component lead wires of 0.002 inch. All this is done with perhaps a requirement for several layers that have to be "drilled." It is, of course, almost a requirement that the laser head be controlled by some automatic process. The use of a computer-controlled numerically-specified programmed machine (CNC) will perform this positioning task readily and accurately. It is of interest to know that in addition to the CO_2 type of lasers, the NdYAG, and Nd:glass lasers are being used in manufacture to "drill", and cut, burr, and machine such materials as plastic, rubber, metal, wood, ceramic, and glass. The applications in manufacture include cutting, drilling, engraving, heat-treatment, soldering, welding, and even insulation removal.

During some welding processes it is necessary to restrict the heat build-up or the material will suffer some thermal distortion or other adverse effect. Because a laser beam can be pulsed at a variable rate, the rate can be set to avoid such thermal problems. Thus there is a definite advantage using lasers in some applications over other types of welding or soldering.

The Coherent Industrial Group, 3210 Porter Drive, Palo Alto, CA 94304 has produced an automated laser welder which is a CO_2 type rated at 575 watts continuous power having an x-y moving optics system controlled by a computer-numerically-controlled machine to generate complex welding patterns. Notice that the use of optics with the laser becomes essential, as we already know it must be. This is called a large-scale type of laser unit and it indicates the possibilities of usage for laser machines of this kind. Of course higher powers from lasers are available. When one uses a laser in a cutting or welding application that requires a high degree of concentration of the laser energy, one must be aware that some factors can effect the performance of the unit. These are the cleanliness of the optical system, the cleanliness of the nozzles, the number of mirrors in the beam path, and the gas pressures if, say, oxygen is added to the "cut" or "weld" to facilitate the cutting or welding. A good clean cut is also a function of the peak power used.

Because a laser beam can be focused to such a small area, it is obvious that it might be used to solder in very tight spaces. The depth of the heat can be precisely controlled and this also means that in wire stripping operations, the insulation can be removed from cables easily and completely and without any damage to the wires, which might occur if a mechanical wire stripper were used. Another advantage of this type of stripping seems evident if one considers a smaller and smaller diameter wire or wiring to be stripped and soldered. It seems that one reaches a point of "diminishing returns" when using mechanical devices. Another idea which persists is that the use and operation of a laser system can be modified by a change of optics—not necessarily a change in the whole laser system itself. As we have pointed out previously, once a laser beam has been generated in a proper frequency and in a proper amount of power, control of that beam is required and this is done through the use of the various "beam control elements" we have previously described. If one plans to go into laser study or operation, then an investigation into "laser optics" is warranted.

RUSSIAN DEVELOPMENTS IN LASER TECHNOLOGY

Some information on this subject is available from Informatics Inc. Rockville, Maryland. Of course, this pertains to the developments some years past, but indicates the direction of the Russian programs in this field. Some information is available on solid-state, and research for liquid and gas lasers, chemical lasers, u-v lasers, non-linear optics, ultra-short pulse operations, and general laser theory.

ABOUT THE FABRY-PEROT INTERFEROMETER

We are well aware of the concept of an interferometer. This

device uses the principle of splitting some signal, ray, or other phenemona and then recombining the two halves after each has been subjected in some way to some outside force or forces. In this way one half becomes a reference against which the other half can be compared, and the changes from the original state or condition can be measured, evaluated and used as a basic to determine the magnitude or phase or effect of the forces that have been influential in changing the other signal.

In the generation of a laser beam from a diode strip laser element, the ends of the element are manufactured so that they are optically "flat" and "parallel." This is so the photons will be able to stimulate others and cause a lasing process. It is this parallel mirror structure, which is called a Fabry-Perot interferometer, and it can be constructed on a larger scale by the proper alignment of two first surface mirrors placed so they are exactly parallel. In the actual interferometer mode, the mirrors are used to separate or combine separate frequencies so that the comparison effect stated can be obtained.

A LASER REMOTE-CONTROL SYSTEM CONCEPT

Of course, if the laser beam as a channel can be used for TV and audio and data communications, then why not use it for some kind of remote control application? There are several good reasons why this would be a good idea. First it seems that one would have the equivalent of a direct wire channel to and from the control site to the controlled object. Second, the channel is relatively immune to outside interference. Third, the equipment does exist for fabricating such a system. One might use any of the conventional "control" code systems over such a link, and so the balance of such a control system might be just as it is now. Only the communications channel is changed and its equipment is added to an existing type of remote-control system. There is no reason why the number of control channels which could be had over such a link would not far exceed what we currently have developed.

What seems to be a problem area is, perhaps, the requirement for some kind of line-of-sight with the remotely controlled object. If we use a light beam, which is what a laser beam amounts to, then we must be able to direct that light beam to the target receiver. Thus such a system might not be usable under some "out-of-sight" situations. Second, haze and fog and other atmospheric conditions might attenuate the transmitted signal so that inadequate reception takes place. Some of these disadvantages might be removed by control of objects from space. It might turn out that there are some situations, which you can possibly imagine, where acoustic or radio frequency or hard wire communications are just not possible or feasible for remote-control applications.

The concept of using a very thin solid-state laser radio-control transmitter to control some object from some distance away is appealing and perhaps practical. The future will embellish this concept. In some experiments done in this area a helium-neon continuous wave laser was used with a receiver that had a 360° light detector antenna. One wonders if it would be possible to maintain control integrity with interference from reflected light signals?

MODULATING A LASER BEAM WITH FM AND AM TYPES OF MODULATION

Acoustical modulation of the amplitude and frequency of a laser beam from a cw helium-neon type of laser has been accomplished. The fact that this modulation can be accomplished with electro-acoustical devices opens a new field of experimentation and development of new types of modulation devices of this kind.

INTEGRATED OPTICAL CIRCUITS

We have integrated electronic circuits. Now we have integrated *optical* circuits. The idea of developing an integrated optical circuit capability for use in a multi-GHz telecommunications application for the military was conceived and put into development. What was desired was a fiber-optic transmission line, multi-terminal multiplexing system through low-loss couplings and suitable modulations. Here, again, we are informed as to developments that will bring about more concrete joining of the technologies of solid-state electronics, fiberoptics, and integrated circuitry. Makes one wonder "where the wires went?"

USING THE PIN DIODE IN A LASER COMMUNICATIONS SYSTEM

An experiment was conducted using lenses and fiber optics so that a curved path could be traversed by a laser beam generated from a pulsed gallium arsenide injection laser when pulsed at about a 10 to 15 kilohertz rate. The PIN diode used in the receiver is a photo-detector. The method of communication is interesting. It is *pulse-position modulation*. This requires excellent stability and synchronization. It is, in fact, the method of radio-controlling so many models that abound in our current world. In this case, instead of each pulse representing a control channel and causing some control (physical) element to move left or right or neutral according to the displacement of that pulse from a nominal or neutral position, the communications concept says that each pulse may follow a sinusoidal variation about its nominal position and thus represent a tone or some data. Each pulse represents a complete communications channel. This is not a new concept but one that has been used for a long time in telemetry applications.

It turns out that the University of Southern California Electrical Engineering Department has made a study of the Pulse-Position Modulation (PPM) as a basis for a communications system using narrow pulses of light. They found that the synchronization requirement was mandatory and has to be very accurate. The use and presence of signal energy in the form of such optical pulses suggested the use of the pulse-edge tracking technique in a binary PPM system to insure the necessary tracking and timing. The timing error variances were shown in this study to be a function of the system's signal-to-noise ratio. This is not an unexpected result of the study. When we consider that individual channels of separated pulses are designated by their timing with respect to some transmitted reference—not to each other—then it is readily apparent that precise timing must be maintained.

It is also apparent that any noise in the system that tends to mask a pulse or reference signal could destroy the synchronization which is the basis for the channeling of the pulses and the amount of excursion of the pulse, which is the data information or communications intelligence.

At the Hughes Research Labs a study was made of the various modulation methods that might be applied to laser-optics systems. Whole systems were considered from the origin of the intelligence through the transmission over optical carriers to detection and display or creation as a signal usable to humans. Modulation systems such as intensity modulation, amplitude modulation, frequency and phase modulation, and both direct and coherent (hetrodyne) reception of signals was considered. One basis for consideration was the information capacity of such systems and the message signal-to-noise ratios.

We learn from some of these studies and experiments that, as in any communications system, the detection of the signals that contain the intelligence are performed with devices that also detect noise in the phenomena used for transmission. It seems impossible to separate the two completely. But it is the *ratio* between what can be used and what cannot be used that becomes very important. It is the basis for many studies and acceptance or rejection of various concepts of intelligence transmission.

In the detection of mode locked (fixed-frequency output) laser signals the passive Fabry-Perot cavity was shown to be a good practical approach to the matched filter optimization for the sensitive detection of the mode-locked signal. Doppler measurements of the relative motion over a wide range of velocities of propagation are possible simply by measuring the cavity length for a peaked output.

We recall that if the length in a cavity is equal to an integral number of half-wavelengths, the cavity will resonate at that frequency. Thus for a long cavity—encompassing an integral number of half-wavelengths for various closely related frequencies, we can readily see how the cavity might resonate at any of the several closely associated frequencies. Given a long cavity length, as is found in a laser stripe-line generating system, one might have a number of frequencies resonating and separated by some amount of the spectrum. Some stripe-line cavity resonators may be as much as 1,000 wavelengths long, and so there is the possibility that some other than the desired frequency will be produced. There may be as many as five to ten modes of operation evident.

AN ANTENNA FOR A 10.6 MICRON PRINTED CIRCUIT

Just consider a printed circuit antenna array for a wavelength of 10.6 microns (10.6 millionths of a meter). That is not very long! But, it was found feasible to develop. The analysis was by the method of moments. An infrared dipole was considered and it was found that with printed circuit technology it had to be a two dimensional type of dipole that had both lateral and axial components of current. The subsequent radiation pattern for such a dipole was analyzed and it was found that a pencil beam that was perpendicular (normal) to the plane of the antenna was capable of production. Of course, solid-state, optical-guide types of antennas also produce such beams. But, it is interesting to know that if we consider the radiation theory concept we could develop an antenna for the smallest of frequencies. As we get higher and higher into the frequency spectrum, however, we begin to require the quantum theory of propagation and analyses based on this theory. The quantum theory is much used when evaluating propagation through various atmospheric conditions.

POTENTIAL USES OF LASERS

Finally, we close this chapter with a kind of quick summary of the possible and potential uses of lasers. We know the usages in medical and biological applications are increasing almost by the hour. Uses in chemical processing and analyses have been indicated in this work. Initiation of controlled thermonuclear reactions may use lasers and we are well aware of the advantages and desirability of using laser beams for communications. The broad bandwidths that are available and the security of the systems make this a very important developmental area. There have been studies that suggest the use of laser beams for study of pollutants in detection and also monitoring of pollution. There are countless applications in the geoscience fields, holographic applications, and in manufacturing and related processes. There will be other uses which are not as yet envisioned, but which *you* might imagine.

Chapter 9

Cable TV Applications of Lasers and Optical Fibers

The day is fast approaching when most homes and businesses will be connected together by fiber-optic cable. When that day arrives we will find that there will be devices available that will permit interactive play between homes, libraries, medical centers, and banks. If we look beyond the social, moral, and artistic concepts of this advent in our lives, then we begin to see a role for lasers and optical fibers. We think of lasers because they operate in those frequency ranges where extremely wide bandwidths can be accommodated. Multiple signals can easily be handled in some manner. Whether the handling will be in a simultaneous mode or a sequential mode will depend on the type of signals and the number to be accommodated. It is not unusual to place many TV channels on a single optical-fiber cable. The advantage of using such a system to convey these signals is that the system is not as subject to electrical interference.

Of course you can imagine the consternation created. Millions of dollars have been spent putting in hard-wire cabling which links homes and offices. One would suppose that companies will not quickly abandon this kind of system to put in the newer and more efficient optical-fiber types of cables. Or will they? The telephone companies, who pass on such costs to their customers, are doing so and one would surmise that this is only after much research and cost-effectiveness analyses and long range operating and usage costs show that the returns are on the positive side of the ledgers. Cable TV may take note of this kind of development.

The demand for cables capable of multi-channel non-interfering signals will continue to grow. With this in mind we now turn our attention to some of the areas of research and development which have been prominent in the field of cable television distribution and operation. These areas point the way to the future kinds of equipment we may become concerned with in theory and practice.

FIBER-OPTIC TRUNK FOR CABLE TV

Each fiber of the cable carries a single video channel. All the components used were either purchased off-the-shelf or were slightly modified versions of stock hardware. The audio signal modulates an audio subcarrier and the combined video and audio signals modulate the LED source. The bandwidth of the combined signals is five megahertz and the frequency deviation of the FM carrier is plus or minus 12 megahertz. At the receiver ends of the cable a silicon avalanche photodiode converts the intensity-modulated optical signal into an FM electronic signal. This is then demodulated in standard form to provide the combined video and audio signals. The systems repeater—needed for any distance of transmission to keep the light levels high enough—consists of a back-to-back combination of one receiver and one transmitter per channel. The optical cable designed for aerial mounting applications was engineered to withstand extreme climatic conditions and it contains a four graded-index fiber interior.

This was developed by Corning Glass with an attenuation of less than 6.5 decibels per kilometer. The cable as designed weighs some 64 kilograms (2.2 lbs equal one kilogram) per kilometer and has a tensile strength of 1,000 newtons. Bandwidth of the system cable is in excess of 500 megahertz-kilometers. (Notice how the bandwidth becomes related to distance in this concept.) The cable was strung on poles that were separated some 78 kilometers. Thus it was proved that such a cable was capable of being supported in an aerial fashion and was strong enough not to break or to prevent use in its primary operational mode. Cable General Corporation was responsible for much of this development.

A WIDE-BAND OPTICAL-FIBER DISTRIBUTION SYSTEM

Using optical fibers in a distribution system requires couplings and connectors and energy propagation as we have previously studied. In a study by Technical University of Denmark, the distribution capabilities and concepts, all types of modulation, capacities, and components were analyzed. The modulation characteristics of the solid-state laser were scrutinized and it was shown in the study that *this laser is unfit* for high quality analog intensity modulation. Analog type transmission can provide only the transmission of a relatively few TV channels at the present required levels of quality, whereas digital types of transmission can accommodate more channels with acceptable quality and a comfortable margin. This study showed that it was not yet possible to simultaneously transmit many TV channels in such a system and that the replacement of a conventional type of TV distribution system with optical fibers cannot yet be accomplished.

However, in another study and an experimental set-up made in Ontario, Canada, a TV supertrunk using optical fibers was installed and was shown to carry 12 color TV signals or channels and 12 FM channels (stereo) over an 8-fiber cable some 7.8 kilometers in length. The system used for transmission was digital and all signals were digitized for multiplexing into a 322 megabit stream that modulated an injection laser diode transmitter. The National Cable Television Association was responsible for a technical paper presented on this subject. Here we find that in a multiplexing concept the necessary signals are transmitted over an optical-fiber cable system. When digital bits are so transmitted, consider that they are serialized or transmitted in a sequential fashion. But, when transmitted fast enough—as these were—then the whole transmission may be considered almost simultaneous as far as equipment and human reception is concerned.

AIR HAZARD PROBLEMS WITH CABLE TV

Another problem which may be solved using optical fibers for cable TV is the known problem of radiation on some aircraft frequencies by existing cable TV systems using hard wire, and electromagnetic waves. As previously pointed out, the use of light frequency not only reduces the possibility of interference but also reduces the possibility of *radiating* electromagnetic signals. This is possibly another way to counter the aircraft interference problems encountered in existing TV cable installations. In 1979 the FCC formed a committee to study the air navigation hazard problem.

USING A HUB OPTICAL-FIBER CABLE TV SYSTEM

Such a system was proposed and it's study indicated that 100 channels could be accommodated using the HUB distribution system and optical-fibers. This system presupposed that pulse-width modulation could be used for ease in switching and that a simple set-top converter could be used to deliver pictures to a standard TV receiver. In this study consideration was given to low-cost connectors which were deemed practical for this application. The study was done by Clifford B. Schrock and Associates, Inc.

USE OF CABLE TV FOR SECURITY PURPOSES

When we see a TV camera in a store or bank we usually do not realize that this is a cable system. Fiber-optical cabling might be very effectively used in security applications simply because it cannot be jammed by radiation devices. Usually the cable "runs" are not long and the reliability and effectiveness of the system must be extremely high. Optical fibers can provide such a system for operating in harsh environments and avoid many of the problems of hard-wire systems. Such systems usually require a motion detector and visual surveillance of areas. The motion detector can alert an operator when he isn't examining a given area at a given time. Other types of sensors can be adapted to work with such a system also.

A LINEAR LASER FOR CABLE TV APPLICATIONS

General Optics Corporation Of South Plainview, N.J. did some research to determine a laser suitable for cable TV applications. They fabricated a laser that would deliver 7 mW of power per face, linear optical power, with respect to the current applied. Their lasers were fabricated from GaAs-GaAlAs wafers grown by the liquid phase epitaxial technique. By carefully controlling the waveguide dimensions with photon implantation, they said they were able to "substantially" improve the linearity. The second and third harmonics were said to be less than 55 dB and 65 dB below the fundamental at 70% modulation. That old devil the signal-to-noise ratio was less than 65 dB. This laser was used to transmit 12 channels of TV signals through a single optical fiber more than a mile in length without any noticeable deterioration of the picture quality. Notice the concept that the laser is an integral part of the optical fiber transmission system.

A LASER BEAM TRANSMISSION USING A
400 MHz BANDWIDTH ELECTRO-OPTICAL MODULATOR

The Japanese have been responsible for the development of

this concept. The optimum configuration of a total-internal-reflection type of laser light beam modulator with surface interdigital electrodes has been obtained. The final specifications for this modulator were an effective half-wavelength voltage of 11.5 volts, a capacitance between electrodes of 9.1 picofarads and a modulation bandwidth of 400 MHz into a proper load impedance. A way had to be found to eliminate the harmful resonances due to acoustic waves excited in the electro-optical crystals by the interdigital electrodes, and this was done. Then a very practical system for transmitting TV signals was built and open-air tested using the modulator. Cross modulation and intermodulation was acceptable.

THE INTERACTIVE OPTICAL-FIBER CATV SYSTEM

We mentioned the advantages of having an interactive cable TV system. This was done in Japan using some 168 subscribers, having a cable length of some 380 km and using a 30 channel capacity network. Thus we find a report that such a system is not only theoretically possible but experimentally practical as well. The results and the equipment used can be found among the IEEE Transactions for the year 1978.

LINEARITY

Notice that through many of these studies the concept of linearity is very important. With digital techniques some non-linearity may be tolerated because equipment can compensate for this, but in the sense it becomes more difficult than one likes to try to compensate for the nonlinearities of light generation from lasers or other suitable optical-fiber light sources. Much work is going on in this field, both in the development of more linear light sources and modulators and such, and also on systems to accommodate nonlinearities. These methods compensate through modified feedbacks and other techniques for the nonlinearities so the output becomes essential linear over the range of interest.

OPTIMIZED DESIGN STUDY FOR A CABLE TV OPTICAL-FIBER SYSTEM

A study was conducted by the Harris Corporation, Melbourne, Florida. This study considered the trade-offs for the application of fiber optics to CATV with the system divided into trunking and local distribution. The performance and cost comparisons for analog-intensity modulation (IM), analog-frequency modulation (FM), and digital (PCM) pulse-code modulation, were considered. The performance and cost of multiplexing for all modulation systems were analyzed. The study concludes that the digital system, multiplexed, with fiber optics is the best for most trunking areas, and then analog approaches are recommended for the distribution on a local basis.

CATV IN EDUCATION

An experimental program was considered in West Germany at the University of Bremen. The authors of the study; Haefner, Issing, and Pruess considered an interactive system of cable TV that could provide two-way communication. They made a proposal to start a pilot program in Germany with 1,000 to 10,000 homes so that reliable data could be collected on the demands of society for such an interactive cable TV program used in an educational mode. This proposal was prepared using the international experience in this area with educational TV, the development of the use of computers in an educational sense and the technological constraints in the development of such a system. The authors actually designed such a system from the technical viewpoint to provide the pilot data.

This type of system, using the latest advances in lasers and optical fibers is forthcoming in the near future. Much has to be considered, not only from the technical viewpoint where simplicity and reliability of technical operation must be paramount, but also the sociological viewpoint as to what, when, and how the educational information must be prepared and presented. A side issue is the implication of "machine" teaching versus "human" teaching in an interactive system, and what is the best mix (if a mix is warranted), of the two methods. This is to be one of the greatest areas of technological and sociological advancement in the not too distant future.

SOME TECHNICAL DATA ON OPTICAL-FIBER TRANSMISSION LINES

Some measurements of optical-fiber waveguides in lengths over 3,000 feet show a total attenuation as low as 2 dB/km and this has been improved on in the latest types of guides. The bandwidth capability of such an optical fiber guide is around 500 MHz and the operating characteristics of these optical-fiber guides are found to be independent of operating frequencies and temperatures. The strength of the optical fiber guide, and the current development of good reliable laser and LED light sources, the ability to join fiber lengths, and to multiplex fibers in junction systems so that feeds to many fibers can be accommodated from just one input fiber, and increasing strength of optical fiber cables made in various forms promises a "snowballing effect" in this technology in the forthcoming years.

MORE ON THE EDUCATIONAL USE OF CABLE TV WITH OPTICAL FIBERS

With satellites making it possible to reach isolated areas, NASA has been participating in experiments along this line. Some material prepared by the U.S. Department of Health Education and Welfare was relayed to many stations in the Rocky Mountain region with two-way voice capability present in the experiment. Although no mention is made of the use of optical fibers or lasers in this experiment, the fact that it was conducted by NASA and others, means that effort has been expended in this area and that this development is growing. It has been stated in some studies that the concept of usage of the educational system with interactive capability will be governed to a large extent by the potential cash flow. Thus the economics of such ventures and experiments and developments is of primary importance.

THE TELEPHONE COMPANY AND OPTICAL FIBERS

Even as we are writing this book we know of at least one "run" which has been made by the telephone company to test out its Picture Phone Concepts. That "run" or link is between Pennsylvania and Washington D.C. Another optical-fiber link is being planned that will link Austin and Dallas Texas. There are others either in the planning or implementation stage in the U.S.

What this brings up is an interesting concept because the telephone linkages already are so vast they link virtually every spot in the world. Our future telephone lines will have every type of information we could possibly desire. Just imagine a range of educational programs from kindergarten to college graduate programs and other special scientific and art programs!

What this can mean to the Cable TV industry is gigantic in its implications. In any event, it seems that you and I as consumers can only benefit no matter which way the development goes. How nice it would be, for example, to be able to access any scientific library in the world just by pushing some buttons! You'd be able to access any stored information anywhere to satisfy your own particular field of interest.

SEMICONDUCTOR DIODE AND SPECIAL LENS

The high radiance GaAs-GaAlAs LED with a self aligning

spherical lens was developed in order to efficiently couple it to a low-loss optical fiber cable. The reason for the coupling was for communications purposes. In the new configuration, the lens is automatically settled into the center of the light-emitting surface in an etched hole containing a clear epoxy resin. When this lens is coupled to a low-loss optical fiber (NA equals 0.14 with a core diameter of 80 micrometers) a coupling efficiency of up to 80% is accomplished. The LED emitting area is about 35 micrometers and the lens some 100 micrometers in diameter. It has a refractive index of 2.0. We note with interest here that this experiment used a special lens which was very small and became a part of the light emitting diode's surface. We also recall that with optical fibers, the special rounding of the ends of such fibers can also produce a lens effect, although this was not used in this case.

A FIBER MOUNTED LED WITH AN OPTICAL-FIBER LENS

The Japanese developed a very small fiber mounted GaAlAs LED to satisfy the requirement for an efficient source of light with some high degree of reliability for an optical fiber communications system. In this case they used the "end effect" of a spherical ended fiber. This means that they rounded the end of the fiber so it produced a lens. In this way the LED-to-optical-fiber coupling was within the range of usability.

MODERN TECHNOLOGY USES A LASER TO MAKE OPTICAL FIBERS

In one conference on optical fibers, the concept of using a CO_2 laser as a primary heat source to "draw" the optical fibers was discussed. It was determined, theoretically, that with a proper selection of parameters a fairly uniform (heat) distribution can be maintained around the cylindrical work-stock. Then an experimental system was constructed using a 90 watt CO_2 laser to draw the optical fibers and proof of this capability was demonstrated. This conference was conducted at the University of Sydney Australia a few years ago.

SAFETY PRECAUTIONS WITH OPTICAL FIBERS

Looking directly into the end of an optical fiber cable connected to a GaAlAs laser or a high radiance LED has been found to be very dangerous. A safety zone has been determined in which the minimum distance from the eye to the end of such a cable or fiber has been determined for an acceptable risk. In the case of a cable with just one "multimode" fiber, *the typical "safe" distance should be at least one meter.* This means you should not look into the end of an optical fiber which is excited by a light source at the opposite end, unless you keep your eye at least one meter away from the end of the fiber. **Actually, we don't recommend that distance. We recommend you don't look into the end of any excited optical fiber—ever.** It was found that with a single mode fiber operation, the critical distance from the eye to the fiber should be at least 4 meters for any safety at all. Of course, considering the type of fiber, the attenuation of the light and such, one *might* not be harmed by closer inspection down to a one meter level, so the study found. **They also found that looking into the end of a cable was just as bad as looking directly at the end of an optical fiber strand. So don't do it!**

With some kinds of light of certain frequencies, we see many photos of experimenters who are wearing goggles of various types working around the light beams. If you must be exposed to laser beams or optical fiber beams, then check what goggles are used—**and use them!**

LASER TO OPTICAL FIBER COUPLING WITH A CYLINDRICAL LENS

This type of lens is not the type usually discussed. It may have some body to it and be shaped on each end so it looks like a cylinder when viewed from the side. Several laser-to-optical-fiber coupling methods have been developed and experimentally checked. A coupling idea using a cylindrical lens is quite simple and effective for enhancing coupling efficiency. As a mounting and aligning structure for this type of laser-to-optical-fiber coupling with a cylindrical lens, a hybrid structure has been proposed using a copper mount on which two vee grooves are produced for aligning all optical elements without adjustment. Even though the mounting structure was produced by a machining method, the average coupling efficiency of about 50% was achieved by using a multimode optical fiber with a half-acceptance-angle of 3 to 4 degrees. These type couplers have been successfully mounted in a hermetically sealed laser package and they seem useful in many practical applications.

SOME MANUFACTURING CONSIDERATIONS FOR OPTICAL FIBERS

An optical fiber waveguide consists of a light-guiding core, optical cladding, and a protective jacket. The core may be dimensioned to a few micrometers to allow only a single, lowest order, mode of propagation. However, a fiber with a much larger core, a multimode waveguide type, offers easier fiber-to-fiber splicing and much better light-coupling efficiency. Low-loss fibers, say with attenuations of 2 to 10 dB/km or better, tend to be expensive and they have small numerical apertures. Thus light-coupling alignments must be pretty exact. High-loss fibers, say 100 dB/km, on the other hand may be low cost and have high numerical apertures. Thus, if one has a choice between a low-loss and a high-loss type, the choice may be more cost-effective if a high-loss type is selected. This is according to a study by Jaeger of Galileo Electric-Optical Co. of Sturbridge, Mass. The choice for "long runs" is obviously a low-loss type he concludes, but for shorter runs, however, the total *system-loss* can actually be greater if a low-loss fiber is used because of the coupling losses caused by the low numerical aperture of the low-loss fibers.

A COMPOSITE OPTICAL-FIBER GUIDE AND STANDARD ANTENNA

Small electromagnetic probes are required in the study of microwave radiation effects on living organisms. To conduct such a study in one case, a miniature isotronic antenna system had an optical fiber waveguide transmission line, or waveguide feed, which was used to couple the data out of the radiation field.

It is a well known fact that in many microwave systems the metallic waveguides may terminate in a "solid" antenna system that is, in effect, a fiber-optic type of waveguide. These antennas, made from such materials as Teflon and other dielectrics, radiate fine beam patterns, while still being of such a shape that they conform to aerodynamic requirements on, say, a military aircraft or even a commercial aircraft.

The reverse now may hold true. Optical-fiber types of waveguides may feed certain types of antennas which do not readily conform to a physical configuration that permits their fabrication out of dielectric materials. Slot antennas may be one type considered. There may be other types that are just as important in their own special applications.

USING OPTICAL DELAY LINES FOR PULSE CODING

At the Illinois Institute of Technology in Chicago, a study was made of the use of optic delay lines to form various kinds of pulses to be used in pulse coding schemes. It was found that optical fiber delay lines offer a simple technique for generating pulse-position codes, or of division of a single output pulse into a series of equally spaced pulses that can then be individually modulated with respect to a time base. With a second set of delay lines in a receiver, it was found possible to receive the original pulse information while sending a lower peak power level. When appropriate low-loss filters are used a considerably longer delay time is possible for the pulses. The study concludes that by using a fiber-delay line type of configuration, a number of signal processing functions analagous to those employed in microwave and computer applications should be realized.

It would seem from this study that one could have some optical fiber lines of various lengths, which are all excited at one end simultaneously. The propagation of the pulses through the fibers will then proceed at some rate according to the fiber type and length. The output ends will, if recombined, then present the initial pulse as a series of pulses, all of which are separated in time according to the time-of-passage through its own optical fiber length. We know that some delays are used in sending pulses through optical fiber windings as, for example, in a laser gyroscope. Thus this concept seems an extension of that proven technique.

TENSILE STRENGTH TESTING FOR OPTICAL FIBER CABLES

At the Stamford Telecommunications Labs a machine that is able to grip the optical fibers in such a way that no damage occurs to the fibers, and then subject that fiber to a tensile strength test was developed. How important this is! One must know how to keep those optical fibers from breaking during installation, environmental changes, and usages. This requires special manufacture of the optical fibers, special cladding materials and methods and a special jacket, which we know is used currently on many optical fiber cables and lines. When considering the use of optical fiber cables and trunking circuits, it is important to know what effect various stresses have on the optical fiber section, and how this may effect the transmission efficiency and accuracy of the data transmitted. The Stanford Study presented some results in this direction. Also, Siemens Research Labs of Munich Germany made many experiments on different glass fibers when they were periodically distorted mechanically. They found a correlation between wavelength and the amount of distortion, and the modes used.

HOW ABOUT A LIQUID CORE OPTICAL FIBER?

Siemens Labs of Munich, Germany conducted experiments with this concept. They found that whereas straight liquid-core fibers have the capability of transporting linearly polarized light without any incident polarization angle, bent filters exhibit some birefrigence with the principle axis fixed to the plane of curvature. The effect of this birefrigence can be compensated for by either periodically altering the plane of curvature by 90 degrees or by winding the fiber into a helix. We learn that a liquid-core optical fiber has been conceived and used and subjected to experiments in light transmission—a fascinating concept!

HOW ABOUT TAPPING INTO AN OPTICAL-FIBER WAVEGUIDE?

This idea has been investigated. If you have an optical fiber waveguide transmitting a certain amount of power, it might be useful to you to be able to tap into that waveguide and extract a given amount of that power for some other purpose. So a method of tapping based upon laterally displacing the waveguides that are butt-joined in such a way that the displacement defines the amount of power transfer that takes place. In other words you just misalign the butt-joint by whatever amount you desire and that governs how much of a light transfer takes place. This is a simple and effective method if you have the means of adjusting the very small fibers or the optical fiber cable. Recall that these fibers are as small as a hair in diameter, and that butt-joining them is a formidable task.

NEAR INFRARED SOURCES IN THE 1 to 1.3 MICROMETER REGION

Low-loss glass optical fiber waveguides are found to be an excellent media for Raman lasers and amplifiers in the near infrared region according to a study at the Bell Telephone Labs at Holmdel, N.J. Wavelength emissions in the 1 to 1.3 micro-meter range is readily available by sufficiently stimulating *Raman scattering* in single-mode silica optical fibers. After experimenting and observing four orders of *Stokes* radiations it was concluded that pulsed, tunable stimulated Raman emission in this wavelength region is possible using kW infrared tunable dye lasers near 1 micrometer as a pumping source.

In another study by the Canadian Department of Communications at Ottawa, Canada, the frequency spectrum of a continuous wave fiber-optic Raman laser was measured and observed to consist of several discrete lines covering the 3 nm spectral range. The output of the laser also contained damped transients that were attributed to strong pump-depletion effects in the filter.

Mentioning the frequency spectrum of optical fibers capability brings to mind the following information that might be useful.

☐ Ultraviolet spectrum
  200 nanometers and below
  200 to 320 nanometers
  320 to 400 nanometers
  400 to 700 nanometers
☐ Infrared spectrum
  0.7 to 1.2 microns
  1.2 to 4 microns
  4 to 15 microns
  15 microns and longer

(The divisions are shown in the ranges that many commercial companies use and much of the technical literature considers these divisions for specific applications.)

A FIBER-OPTIC DIGITAL DATA AND TV LINK

Mentioned recently is the fact that optical fibers might carry more than one TV program. The programs might be transmitted in a simultaneous manner using a digital code. In fact, this is said to be the planning stage for the Summer Olympics in Los Angeles for 1984. Bell Laboratories at Holmdel, N.J. are working hard to finalize such a system to enable the transmission of acceptable TV signals over a fiber-optic linkage using digital modulation. It is currently planned that this system will receive data from as many as 27 sources, digitize it, and transmit it over the optical-fiber cables at a 90 megabits/second rate. The transmission is to take place over the Pacific Telephone's 330 route-miles of optical fibers now being installed as part of this company's "West Corridor Project."

In order to make such a scheme workable it is planned to use a multiplexing concept based on wavelengths and division of the wavelengths. This means that the signals will be digital "bits" modulated on three wavelengths. One will be at 0.82 micrometers, one at 0.87 micrometers, and one at 1.31 micrometers. All three will be sent simultaneously over a single optical fiber.

In communications techniques there is a concept called "data compression", which means squeezing information into a smaller frequency range than it normally occupies. In the concept developed by Bell Telephone, it is planned to use this technique to "squeeze" nine bits of information into an eight-bit code which seems to be the maximum code-word length that an optical fiber can handle reasonably well and maintain the quality required for the transmission of the data. If a nine-bit analog-to-digital data link were used, it is assumed that the optical fiber link could not handle the high-speed data rate involved without other problems developing.

A LASER READER FOR HOME RECORDERS

Actually the laser doesn't read the information. It supplies the light source and then there are optics and optical fibers and photodiodes to help convert the reflected light pulses back into digital signals. These are converted into analog signals by digital-to-analog techniques and amplified and presented to disply or audible output devices.

It is said that such a digital system is being developed in Japan by Matsushita and will have a stereo signal consisting of a 16-bit code sampled at a 44.056 kHz rate on 12 tracks on each side of a cassette for a total of 24 tracks. The prototype system which Masushita has produced has micro-cassettes and it seems to work fine except that it uses a slightly smaller code. It uses 14 instead of 16 bits and the sampling is at a reduced rate of some 33 kHz. It is said that the dynamic range is almost equal to 85 kHz. Lasers used in this type application must be small, reliable, and arranged so that they can track the microgroves that provide the reflective pulses needed to complete the system's operation. One would imagine that in the first stages of such a system, where high sensitivity is required, that the use of optical fibers (immune to electrical noises) might be used.

THE RAMAN EFFECT

This is a characteristic of light first observed by C. V. Raman in 1928. In essence it means the scattering of light from a gas, liquid, or solid with a shift in wavelength in the incident radiation. This effect is also called the Smekal-Raman effect. Smekal did some theoretical work and theorized about some vibrational effects of molecules when subjected to some incident radiation. The classical theory indicates that the incident radiation, when so scattered, should not be changed. Smekal theorized that the effects of scattering and radiation could be explained by quantum-mechanical theories, and so they were, Certain effects could be observed experimentally (the scattered frequencies related to Stokes and anti-Stokes lines). When the most intense radiations produced by scattering are unchanged this is known as Rayleigh scattering.

TESTING OPTICAL FIBERS

How do you test an optical fiber or optical-fiber cable? You need to know the attenuation per kilometer, and the purity and frequency of the source.

An instrument for the nondestructive measurement of attenuation in optical fibers was developed by the Stard Telecommunications Labs. Ltd., of Harlow, Essex, England for checking the laser in the system and the light detector equipment. This instrument was the first of a series designed to cover all tests required for optical fibers and it incorporated its own gallium-aluminum-arsenide (GaAlAs) laser source. The attenuation of optical fibers with some range of core sizes, and whose losses might range as high as 60 dB, may be measured within 1 dB accuracy with the instrument. The method used is the well known one of "electrical substitution." A comparison is made with a short (zero dB loss) section of similar fiber. Then the system detector can be checked for a go no-go kind of performance. Also the detector can be calibrated to check the system's laser source.

In all of the new installations of optical fibers, the method of testing and checking for the three very important system elements: the light source, the optical fiber cable or strand, and the detection system, will be mandatory. Not only will the light source have to be tested for power output capability, but also for generation of the proper light frequencies and polarities (probably) and monochromaticity and mode. Instrumentation for detection of faults and flaws and location of abberations will be required and designed, and expert and intelligent personnel will be required to operate such instruments.

A study was made by the Italians that involved the determination of the error probability in digital transmission over optical fibers. The method took into account the principal phenomena that influences the error probabilities. First an approximate expression is obtained for the error probability, based on simplifying assumptions and hypothesis. Then both the exact procedure and approximate expression are applied to a particular case in which the amplifier-equalizer has a transfer function derived from the hypothesis that the isolated impulses both at the output and at the input of the receiver should have a Gaussian shape.

It is easy to see that all theoretical and practical methods available are being used to determine the capability of the optical fiber transmission systems concepts. When finally one has all the equations and instrumentation required to design and test such equipment then optical-fiber system design will take its place alongside microwave systems and hard-wire systems.

GRADED INDEX FIBERS

The requirement in many applications is to have optical fibers with a *Graded index of refraction.* We already know what the index of refraction means, so now we are considering how to make a fiber that will have different gradations of the index of refraction as we progress outward from the center of the fiber core. It has been possible to make such a fiber by using a rod of a lower index of refraction and then cause it to shrink onto a rod of a higher index of refraction. Thus the index of refraction gets smaller in a step-wise fashion as we proceed out from the core center.

In the "old days" of guided missiles, a system was proposed using a Luneberg lens as a focusing element for radar waves both in emission and returning echoes. The idea in this lens was that it would be round (like a ball) and have a variable index of refraction for microwaves. The index of refraction was changed by the use of some rods of various sizes in the ball's construction. Also, it turns out that no matter from what direction the microwaves entered the lens, they would come to a focus at a point (or small area) directly opposite the direction the rays entered the ball. That was why it was called a lens.

In a general way we can just consider that optical fibers can be made so that the index of refraction varies either in a gradual manner, from core center to the cladding, or in a step manner where the index may change significantly over a very small spatial change going from, say, one type of glass abruptly to another kind of glass in the fiber. It is known that the bandwidths of graded fibers are very much larger than can be obtained with step-type fibers but the numerical aperture for a graded fiber is less than that for a step-type fiber. This means a more accurate positioning of the light source (to get the light rays into the fiber) is required with the graded fiber types. It is also known that the attenuation of the light rays in a graded fiber is less than it is for a step-index type of fiber. Thus, when we are aware that the Nippon Electric Company of Japan is doing research on digital transmission of a high-speed nature over optical fibers of a graded-index type, we have some understanding what is meant concerning the type of fiber and why that fiber might be used. It is because of the wide bandwidths needed for this type of transmission.

SOME DATA ON AN EXPERIMENTAL
OPTICAL-FIBER SYSTEM INTEROFFICE TRUNK

The basic elements of such a system are a light source, a means for modulating that light source, a transmission system (optical fibers) and a receiver to convert the light pulses (assuming digital transmission which is most predominant currently) into electrical signals, and the shaping and reconversion of the electrical signals into some usable form. In this experimental operation the laser light source used was a GaAlAs injection type of laser. Avalanche photodiodes were used as receivers. The light source from the laser was electronically monitored, controlled, and modulated. The gain of the receiver was controlled by varying its bias supply. We note here that an avalanche photodiode in a typical circuit may use a bias supply that can be so set that it gives a threshold value of detection for incoming light pulses. A high bias would cause an intense signal to be present before it was recognized and a threshold value of bias would result in the utmost sensitivity for the detector. You might want to file away this information if you do experiments with lasers, photodiodes, and optical fibers.

SOME PRINCIPLES OF OPTICAL-FIBER CABLE DESIGN

The problem of making optical fiber cables was studied by the Stan Telecom Labs Ltd. at Essex, England, and they came up with some design criteria which seems appropriate. In their study the main concerns were brittleness of the optical fiber material, the optical sensitivity, and the mechanics of manufacture, installation, and servicing of such cables. The necessary mechanical strengths and properties for the cabling of the optical fiber light-guides was obtained by the application of plastic coatings. A number of such coated fibers may then be combined with a base cable "stringer" or strengthener that increases the total strength of the assemblage. This is done with such fillers and wrappings as required to maintain the geometric register. The unit may then be plastic wrapped and it can be covered with water barriers and given even more strengthening members so a large tensile strength can be obtained. Many such optical-fiber sections may then be combined into complete cables that have a very high capacity for information transmission. The basic design requires that the fibers be protected against tensile and radial stresses, bending conditions, and environmental problems. If you have ever examined the optical fibers used in "light displays" at novelty stores and have taken one such fiber out of the display to examine it, you'll find that it is very small and very brittle—it breaks easily. Of course, these aren't the fibers used in cables, but this does give us some idea of the problems associated with making a good strong optical-fiber cable that can be handled in the field like any other cable.

LEAKY FIBERS

In some of the literature on optical fibers we find that there is such a thing as a "leaky" fiber. We assume it to be the loss of the light rays of some given frequencies because these rays propagate at some point at more than the "critical" Brewsters angle and so escape into the cladding and are absorbed or attenuated very drastically.

We know of "leaky" microwave systems, some of which are designed intentionally to "leak" radiation. We know of situations where the microwave energy "gets out of" a waveguide at a joint or a twist or some other discontinuity in the waveguide structure. So it is with light rays, which as we have learned can, in many ways, be treated as just smaller wavelength electromagnetic radiations. Thus, anything that would provide any kind of discontinuity or imperfection in the fabrication of an optical fiber might cause the optical fiber to "leak" energy at that point. Depending on the frequency and mode(s) of operation of the optical-fiber guide, some frequencies may "leak" and some may not. In some cases it has been found that undesirable radiations can be attenuated by this "leaky process" and this is good because it reinforces the desired radiations.

DETERMINING THE TRANSFER
FUNCTION FOR AN OPTICAL-FIBER GUIDE

Graduate students (and some undergraduate students) who have calculators and computers at hand may like to try to determine the mathematical transfer function for an optical fiber light guide. this is how it was done in Germany: A short laser pulse of less than 100 picoseconds is coupled into the fiber and the optical pulse appearing at the output of the fiber is recorded. The complex transfer function is then calculated from the two optical pulses (input and output) with the aid of the Fourier transform. This technique is said to be usable for many different types of fibers.

THE LIFE-SPAN OF INJECTION LASERS AND LEDS

Some studies and experiments at the Princeton, N.J. RCA Labs indicate that some lasers, GaAlAs types, may have a very long lifetime for useful outputs. In some tests these lasers have been operated for longer than 10,000 hours and the study, which was conducted some time ago, indicates that LEDs and cw lasers may have a useful life-span of over 100,000 hours. This becomes very important when we consider small size lasers and LEDs being used as an integral part of an optical-fiber cable transmission system.

LEARNING ABOUT OPTICAL WAVEGUIDES

Information might be obtained from the University of Virginia at Charlottesville concerning a program and equipment that was studied for the teaching of optical-fiber concepts. In May of 1977 they presented a paper in which they described an inexpensive integrated optics laboratory for teaching and research. The paper lists and specifies the basic materials needed to fabricate optical waveguides and make guide measurements. Several methods of fabricating optical waveguides are discussed, and the dipping method used with glass substrates and polymer solution were considered. Polystyrene waveguides were used to demonstrate some of the fundamental concepts of guided waves, propagation modes, coupling, and attenuation.

SOME RUSSIAN MANUFACTURING METHODS OF OPTICAL FIBERS

So important is the concept of using optical fibers in all manner of communications and data transmissions that most countries of the world are engaged in research on this subject. In a paper from The Institute of Radio Engineering and Electronics Academy of Sciences, Moscow, some early manufacturing methods of preparing optical fibers was stated as follows: "Low-loss optical fibers were prepared as follows: Halide vapors were oxidized inside a quartz jacket tube placed in a traveling hot zone. The tube was then collapsed into a rod blank and fibers were drawn (from it). The deposited halide films were doped with boron and phosphorous in accordance with selected programs. The losses were about 4 dB/km."

SERIAL-ACCESS COUPLERS AND DUPLEXES FOR OPTICAL FIBERS

It has been found that these can be obtained for single-strand multi-mode types of optical waveguides by using conventional optical components such as beam splitters and lenses. This approach, however, is difficult because of the small optical fiber diameters and the extremely close tolerances required during fabrication and assembly. Some alternatives are the taper fiber coupling and the welded access (T) coupler. In single-strand fully duplex systems, the amount of light reversed, or backscattered, as the optical pulse travels down the fiber is an important consideration. The tapered

308

fiber coupler has been demonstrated as a convenient means to experimentally measure this "backscattering effect."

A RADIATION-HARDENED OPTICAL-FIBER SYSTEM

The Harry Diamond Labs at Adelphi, Maryland developed a radiation hardened optical fiber transmission system with a 400 MHz bandwidth and linear response. An amplitude modulated cw laser diode is used in the transmitter with automatic bias compensation applied instead of temperature control. This system can select, via an optical-fiber control link, the attenuation between the transmitter and the receiver. The transmitter and optical-fiber portions of the system were hardened to withstand a photon irradiation of more than 5×10^5 rads from a Betatron.

COUPLING A LASER INTO AN OPTICAL FIBER

Two methods of coupling a multimode fiber to a GaAs double heterostructure laser were investigated. First, direct coupling into a plane-fiber-end-face, and second, coupling into a cylindrical glass-fiber lens. The power coupled was measured for three different lasers and one laser was adjusted by its bias current. Some optical feedback was observed when the coupling between the laser and the optical fiber end varied just a few tenths of a micron. The tolerances for the fiber end type of coupling are, as expected, very tight.

A TUNABLE RAMAN OPTICAL FIBER LASER

From Ottawa, Canada's communications research center, comes a note that they have shown that continuous tuning of the oscillation frequency of a continuous wave Raman optical-fiber laser was possible. The device they used operated over a 3 nm band in the visible light range and exhibited low threshold and high conversion efficiency.

MULTIMODE AND SINGLE MODE

What these mean, in essence, is that the optical fiber may have only one frequency propagated through it (single mode) or it may have many different frequencies propagated through it (multimode). If a pulse is propagated in multimode form, we find that some frequencies may propagate faster than others and so there is a widening of the pulse. It seems that the refractive index governs, to a large extent, this effect. Stepped or graded optical fibers may help to avoid this problem.

OPTICAL-FIBER LOSSES

Some of the things which cause losses in energy when light is transmitted through optical fibers are: the Rayleigh scattering, the losses due to tunneling from the modes closest to cut-off into the cladding, and the forward scattering caused by imperfections in the optical fibers that are large compared to the wavelengths transmitted.

THINKING IT OVER

As we think back through the past pages, we realize that a new era in communications and computer techniques is upon us. In years past it was predicted that we would someday have "solid blocks" from which we could get sound, music, and and by looking at the block we could see various images. It seems that such a day is at hand. The solid-state devices such as we have considered in this text approach that "solid block" condition of electronics.

No longer will we be concerned with capacitors, resistors, wiring, and soldering. Everything electronic will one day become a chemically and atomically bonded solid and do miraculous things for us. Computers will shrink to the size of a playing die and be able to access other computers and memory banks everywhere through means we cannot now envision. Who knows, perhaps someday in the future will come the means by which we can learn in one year what we learn in 20 years today! How strange it will seem to wake up some morning, after setting the "educational knowledge brain-wave organizer" the night before, and find ourselves a dentist, doctor, engineer, physicist, aeronautical engineer, or whatever! Those little die cubes will have accessed stores of knowledge everywhere and caused that information to be implanted in our own changed brain contours and patterns!

The miracle of optical fibers will permit more information, intelligence, and data to be transferred from one point to another more quickly and precisely than we'd ever thought possible, and as their algorithms and equations are developed and proved fiber optics shall become as wires, easy to construct to precise tolerances, accurate and indefatigable in operation, and a type of "forever" circuit which will enhance our lives and ways of living.

It is good for us to know about fiber optics and to understand them even though we may not experiment with them. We may not be able to get fiber optics and all the necessary experimental equipment. But, time will solve this latter inconvenience. Strands of optical fibers used in various capacities (from kits for the younger experimenters, to various devices using optical fibers for the more advanced) will become very common in the future. More and more uses for these solid-state devices will become available to us.

Meantime read, study, learn and experiment as best you can, while looking forward to the day when wiring as we know it might become almost obsolete in the communications and data transmission sense. Then, the knowledge already gained in optical fibers, light sources, receivers, and light control elements will stand you in good stead for advancement in the quickest and most practical manner into the science and engineering of the future. Always remember that the future is not ten or twenty years from now. The future begins tomorrow!

Glossary

This glossary is courtesy of The National Telecommunications and Information Administration, Boulder, Colorado.

angular misalignment loss—An optical power loss caused by angular deviation from the optimum alignment of source to optical waveguide, waveguide to waveguide, or waveguide to detector. See also: extrinsic junction loss; gap loss; intrinsic junction loss; lateral offset loss.

antireflection coating—A thin, dielectric or metallic film (or several such films) applied to an optical surface to reduce the reflectance and to increase the transmittance. Note: The ideal value of the refractive index of a single layered film is the square root of the product of the refractive indices on either side of the surface to which it is applied. The ideal optical thickness being one quarter of a wavelength. See also: dichroic filter; Fresnel reflection; reflectance; reflection loss; transmittance.

APD—Abbreviation for avalanche photodiode.

attenuation—The diminution of signal amplitude or power. In optical waveguides, attenuation results from several mechanisms that may operate simultaneously: absorption; scattering; and losses into radiation modes. Attenuation is generally expressed in dB/km, assuming approximate uniformity with length. See also: coupling loss; differential modal attenuation; equilibrium mode distribution; extrinsic junction loss; insertion loss; intrinsic junction loss; leaky modes; macrobend loss; material absorption; material scattering; microbend loss; Rayleigh scattering; spectral window; transmission loss; waveguide scattering.

attenuation coefficient—The normalized rate of change of total optical power with respect to distance along the waveguide. It is defined by the equation

$$P(z) = P(0) \ 10^{-\left(\frac{\alpha z}{10}\right)}$$

where P(z) is the power at a distance (z) along the guide and P(0) is the power at z=0; (α) is in dB/km if (z) is in km. From this equation,

$$\alpha z = -10 \ \log_{10} \left[\frac{P(z)}{P(0)}\right]$$

This assumes that (α) is dependent of (z); if otherwise, the definition must be given in terms of incremental attenuations as:

$$P(z) = P(0) \ 10^{-\int_0^z \frac{\alpha(z)}{10} \ dz}$$

attenuation limited operation—The condition prevailing when the signal amplitude (rather than distortion) limits the communication capacity. See also: bandwidth limited operation; distortion limited operation.

avalanche photodiode (APD)—A photodiode designed to take advantage of avalanche multiplication of photocurrent.
Note: As the reverse-bias voltage approaches the breakdown voltage, hole-electron pairs created by absorbed photons acquire sufficient energy to create additional hole-electron pairs when they collide with ions; thus a multiplication (signal gain) effect if achieved. See also: photodiode; PIN photodiode.

axial ray—A light ray that travels along the optical axis. See also: geometric optics; meridional ray; optical axis; paraxial ray; skew ray.

backscattering—The scattering of power into the direction opposite that of signal transmission.

bandpass filter—See optical filter

bandwidth—A continuous range of frequencies between a lower and upper limit.

bandwidth limited operation—The condition prevailing when the frequency spectrum or bandwidth, rather than the amplitude (or power) of the signal, is the limiting factor in communication capability. The condition is reached when the system distorts the shape of the waveform beyond tolerable limits. For linear systems, bandwidth limited operation is equivalent to distortion limited operation. See also: attenuation limited operation; distortion limited operation; linear optical element or system.

barrier layer— In an optical waveguide, a localized minimum in refractive index in the region of the core/cladding interface.

baseband response function—Synonym for transfer function.

beam diameter—The diameter of a circle, concentric with a beam, through which passes a specified fraction of the total power in the beam. The term "beam diameter" is only useful for beams that are or are assumed to be circular in cross section.

beam divergence angle—Half the vertex angle of that cone that encompasses a circle of diameter equal to the beam diameter at all points in the far field. Beam divergence angle is strictly only meaningful when describing the far field of beams that are or that are assumed to be circular in cross section. See also: beam spread; beamwidth; collimation; far field.

beamsplitter—A device for dividing an optical beam into two or more separate beams.

beam spread—The angle between two planes, within which passes a specified fraction of the total power in an optical beam. Beam spread is generally specified for two orthogonal orientations. Note: This definition is useful in characterizing beams that are not circular in cross section. See also; beam divergence angle.

beamwidth—The linear dimension of the region over which the beam irradiance falls within specified limits. See also: beam divergence angle; irradiance.

bidirectional transmission—Signal transmission in both directions along an optical waveguide or other component.

birefringent medium—A material that exhibits different indices of refraction for orthogonal linear polarizations.

blackbody—A totally absorbing body (that reflects no radiation). Note: In thermal equilibrium, a blackbody absorbs and radiates at the same rate; the radiation will just equal absorption when thermal equilibrium is maintained. See also; emissivity.

Boltzmann's constant—A constant that relates the average energy of a molecule to the absolute temperature of the environment. Defined as 1.38×10^{-23} joules/K. K=Kelvin.

boule—In the manufacture of optical fibers, a synonym for Preform.

bound mode—A propagating mode whose power is predominantly in the core of the waveguide. Synonyms: Core mode; Guided mode. See also: cladding mode; leaky mode; mode; normalized frequency; unbound mode.

Brewster's angle—For light incident on a plane boundary between two regions having different refractive indices, that angle of incidence at which the reflectivity of light having its electric field vector in the plane defined by the direction of propagation and the normal to the surface is zero. For propagation from medium (1) to medium (2), Brewster's angle is given by:

$$\theta = \text{arc tan } \frac{N_2}{N_1} \quad N = \text{index of refraction}$$

brightness—An attribute of visual perception, in accordance with which a source appears to emit more or less light: obsolete. See also; radiance; radiometry.

Note 1: Usage should be restricted to nonquantitative statements in reference to physiological sensations and perceptions of light.

Note 2: "Brightness" was formerly used as a synonym for the photometric term "luminance" and (incorrectly) for the radiometric term "radiance".

buffer—See fiber buffer.

bundle—See fiber bundle.

cable—See optical cable.

cable assembly—See multifiber cable; optical cable assembly.

cavity—See optical cavity.

chemical vapor deposition (CVD) technique—A method of fabricating optical waveguide preforms by causing vapors to react and form a deposit that may be in the form of glass particles. See also: double crucible technique; ion exchange technique; preform; rod-in-tube technique; soot technique; vapor phase axial deposition technique.

chirping—A rapid change (as opposed to long term drift) in the emission wavelength of an optical source. Chirping is most often observed in pulsed operation of a source.

chromatic dispersion—Redundant synonym for dispersion.

cladding—In an optical waveguide, a homogeneous dielectric that concentrically surrounds the core with a lower refractive index. See also; core; normalized frequency; optical waveguide.

cladding diameter—Specified through the use of a tolerance field. See also: core diameter; tolerance fields.

cladding mode—A propagating mode whose power is predominantly in the cladding. See also: bound mode; leaky mode; mode; unbound mode.

cladding mode stripper—A device that employs a material having an index approximately equal to (or greater than) that of the waveguide cladding which, when applied to the waveguide, provides escape for cladding modes.

coherence area—The area in a plane perpendicular to the direction of propagation over which light may be considered highly coherent. Commonly the coherence area is the area over which the degree of coherence exceeds (0.88).
Note: Light is considered highly coherent when the degree of coherence exceeds (0.88), partially coherent for values less than (0.88), and incoherent for "very small" values. See also: coherent; degree of coherence.

coherence length—The propagation distance over which a light beam may be considered coherent. If the spectral linewidth of the source is ($\Delta\lambda$) and the central wavelength is (λ_o), the coherence length in a medium of refractive index (n) is approximately $\lambda_o \frac{2}{N\Delta\lambda}$. See also: degree of coherence; spectral linewidth.

coherence time—Coherence length divided by the phase velocity of light in a medium; approximately given by ($\lambda_o \frac{2}{C\Delta\lambda}$). See also: coherence length; phase velocity.

coherent—Characterized by a fixed phase relationship among points on an electromagnetic wave.
Note: A perfectly monochromatic wave would be perfectly coherent at all points in space. In practice, however, real waves are categorized according to their spatial coherence or their temporal (phase) coherence: A wave has a high degree of spatial coherence if the radiation is coherent at points distributed on a wavefront (or on a plane approximately normal to the direction of propagation). Likewise, a wave has high temporal coherence if the radiation is coherent at points distributed approximately along the direction of propagation. See also: coherence area; degree of coherence; monochromatic.

coherent bundle—Synonym for aligned bundle.

coherent radiation—See coherent.

collimation—The process by which a divergent or convergent beam of radiation is converted into a beam with the minimum divergence possible for that system (ideally, a parallel bundle of rays). See also: beam divergence angle.

connector—See optical waveguide connector.

connector insertion loss—See insertion loss.

conservation of radiance—A basic principle stating that no passive optical system can increase the quantity (LN^{-2}) where (L) is the radiance of a beam and (n) is the local refractive index. Formerly called "conservation of brightness" or the "brightness theorem." Synonym: radiance theorem. See also: brightness; radiance.

core—The center region of an optical waveguide through which is intended that light be transmitted. See also: cladding; normalized frequency; optical waveguide.

core diameter—The diameter of the waveguide at the point where the refractive index of the core exceeds that of the cladding by (k) times the difference between the maximum refractive index in the core and the minimum refractive index in the cladding, where (k) is a specified constant $[0 < K < 1]$ specified through the use of a tolerance field. See also: cladding; core; tolerance fields.

core mode—Synonym for bound mode.

cosine emission law—Synonym for Lambert's cosine law.

coupler—See optical waveguide coupler.

coupling—See mode coupling.

coupling loss—The power loss suffered when coupling light from one optical device to another. See also: angular misalignment loss; extrinsic junction loss; gap misalignment loss; insertion loss; lateral offset loss; intrinsic junction loss.

critical angle—When light propagates in a homogeneous medium of relatively high refractive index (n high) toward a planar interface with a homogeneous material of lower index (n low), the critical angle of incidence is defined by:

$$\theta_c = \sin^{-1}\left[\frac{N_{low}}{N_{high}}\right]$$

Note: When the angle of incidence exceeds the critical angle, the light is totally reflected by the interface. This is termed total internal reflection. See also: acceptance angle; reflection; refractive index; step index profile; total internal reflection.

current—See driving current—threshold current.

curvature loss—Synonym for macrobend loss.

cut-off frequency—See single mode optical waveguide.

cvd—Abbreviation for chemical vapor deposition.

D*—(pronounced "D-star") A figure of merit often used to characterize detector performance, defined as the reciprocal of noise equivalent power (NEP), normalized to unit area and unit bandwidth.

$$D^* = \frac{\sqrt{A\ (\Delta f)}}{NEP}$$

where (A) is the area of the photosensitive region of the detector and (Δf) is the bandwidth of the modulation frequency of the incident radiation. Synonym: specific detectivity. See also: detectivity; noise equivalent power.

dark current—The external current, under specified biasing conditions, that flows in photosensitive detectors when there is no incident radiation.

degenerate waveguide modes—Waveguide modes that have either the same phase or group velocity. See also: group degenerate modes; group velocity; mode; phase degenerate modes; phase velocity.

degree of coherence—A measure of the coherence of a light source; the magnitude of the degree of coherence may be shown to be equal to the visibility of the fringes of a two-beam interference experiment, where:

$$V = \left[\frac{I_{max} - I_{min}}{I_{max} + I_{min}}\right]$$

(I_{max}) is the intensity at a maximum of the interference pattern, and (I_{min}) is the intensity at a minimum.

Note: Light is considered highly coherent when the degree of coherence exceeds (0.88), partially coherent for values less than (0.88), and incoherent for "very small" values. See also: coherence area; coherent; coherence length; interference.

density—See optical density.

detectivity—The reciprocal of noise equivalent power (NEP). See also: D*; Noise equivalent power.

dichroic filter—An optical filter that transmits light selectively according to wavelength (most often, a high-pass or low-pass filter). See also: optical filter.

dichroic mirror—A mirror that reflects light selectively according to wavelength.

differential modal attenuation—The differences in attenuation among modes.

differential modal delay—The differences in propagation delays among modes owing to their differing group velocities. Synonym: Multimode group delay. See also: group velocity; mode; multimode distortion.

differential quantum efficiency—The slope of the light-current curve, used to describe devices that have nonlinear output-input characteristics.

diffraction—The deviation of a wavefront from the path predicted by geometric optics when a wavefront is restricted by an opening or an edge of an object.

Note: Diffraction occurs whenever a beam of light is restricted in any way and may still be important when the opening is many orders of magnitude larger than the wavelength. Often distinguished from intereference by its pattern.

diffraction limited—A beam of light is diffraction limited if: a) the far field beam divergence is equal to that predicted by diffraction theory, or b) in focusing optics, the impulse response or resolution limit is equal to that predicted by diffraction theory.

diffuse reflection—See reflection.

diode laser—Synonym for injection laser diode.

dispersion—A term used to describe the chromatic or wavelength dependence of a parameter. The term can be used, for example, to describe the process by which an electromagnetic signal is distorted because the various frequency (i.e., wavelength) components of that signal have different propagation charactersitics. The term is also used to describe the relationship between refractive index and wavelength.

Note: Signal distortion in an optical waveguide is caused by several dispersive mechanisms; waveguide dispersion, material dispersion, and profile dispersion. In addition, the signal suffers degradation from multimode "distortion," which is often (erroneously) referred to as multimode "dispersion." See also: distortion; intramodal distortion; material dispersion; material dispersion parameter; multimode distortion; profile dispersion; profile dispersion parameter; waveguide dispersion.

distortion—An undesirable change of signal waveform shape.

Note: Signal distortion in an optical waveguide is caused by several dispersive mechanisms: waveguide dispersion, material dispersion, and profile dispersion. In addition, the signal suffers degradation from intramodal distortion and multimode "distortion" which is often (erroneously) referred to as multimode "dispersion." See also: dispersion: intramodal distortion; multimode distortion.

distortion limited operation—The condition prevailing when the shape of the signal, rather than its amplitude (or Power), is the limiting factor in communication capability. The condition is reached when the system distorts the shape of the waveform beyond tolerable limits. For linear systems, distortion limited operation is equivalent to bandwidth limited operation. See also: attenuation limited operation; bandwidth limited operation; distortion; multimode distortion.

divergence—See beam divergence angle.

double crucible technique—A method of fabricating an optical waveguide by melting core and clad glasses in two suitably joined concentric crucibles and then drawing a fiber from the combined melted glass. See also: chemical vapor deposition technique; rod in-tube technique; soot technique; vapor phase axial deposition technique.

driving current—The electrical input current that drives a semiconductor light source. See also: lasing threshold; threshold current.

efficiency—See differential quantum efficiency; power efficiency; quantum efficiency.

electric vector—The electric field vector associated with a light wave. The electric field vector direction specifies the polarization and its length, the amplitude of the electric field.

electroluminescence—Nonthermal conversion of electrical energy into light in a liquid or solid substance. One example involves the photon emission resulting from electron-hole recombination in a (pn) junction. This is the mechanism involved in the injection laser. See also: injection laser diode.

electro-optic effect—A change in a material's refractive index under the influence of an electric field.
Note 1: *Pockels and Kerr effects are electro-optic effects that are respectively linear and quadratic in the electric field strength.*
Note 2: "Electro-optic" is often erroneously used as a synonym for "optoelectronic." See also: optoelectronic.

emissivity—The ratio of power radiated by a substance to the power radiated by a blackbody at the same temperature. Emmissivity is a function of wavelength and temperature. See also: blackbody.

equilibrium condition—Synonym for equilibrium mode. Distribution.

equilibrium coupling length—Synonym for equilibrium length.

equilibrium length—The length of multimode optical waveguide, excited in a specified manner, necessary to attain the equilibrium mode distribution. Synonyms: equilibrium coupling length; equilibrium mode distribution length. See also: equilibrium mode distribution; mode coupling; nonequilibrium mode distribution.

equilibrium mode distribution—The distribution of power among the modes after transmission through a requisite length of multimode optical waveguide such that thereafter (in distance) the relative power distribution among the various modes remains constant.

Note: The requisite length, which typically varies from several hundred meters to a few kilometers, is dependent upon; various parameters of the waveguide; wavelength; and initial launching conditions. Synonyms: equilibrium condition; steady state condition. See also: equilibrium length; mode; mode coupling; nonequilibrium mode distribution.

equilibrium mode distribution length—Synonym for equilibrium length.

equilibrium mode simulator (EMS)—A device used to create an approximation of the equilibrium mode distribution. This distribution may be achieved by using selective mode excitation or mode filters either with or without mode scramblers. See also: equilibrium mode distribution.

equilibrium radiation angle—The radiation angle of an optical waveguide having an equilibrium mode distribution. See also: acceptance angle; equilibrium mode distribution; numerical aperture; radiation angle; radiation pattern.

equilibrium radiation pattern—The output radiation pattern of an optical waveguide having an equilibrium mode distribution, measured as a function of angle or distance from the waveguide axis.

Note 1: Far field equilibrium radiation pattern is measured as a function of angle. Near field equilibrium radiation pattern is measured as a function of distance from the waveguide axis.

Note 2: The equilibrium radiation pattern is independent of waveguide length—beyond the equilibrium length—and excitation conditions but may be a function of wavelength. See also: acceptance angle; equilibrium mode distribution; equilibrium radiation angle; numerical aperture; radiation angle; radiation pattern.

evanescent field—A time varying electromagnetic field whose amplitude decreases monotonically but without an accompanying phase shift in a particular direction is said to be evanescent in that direction.

extrinsic junction loss—Those junction losses that are caused by different geometric and optical parameter mismatches when two nonidentical optical waveguides are joined. See also: angular misalignment loss; gap loss; intrinsic junction loss; lateral offset loss.

far field diffraction pattern—The diffraction pattern of a source (including the output end of an optical waveguide) observed at an infinite distance from the source. Theoretically, a far field pattern exists at distances that are large compared with $(S \frac{2}{\lambda})$, where (s) is a characteristic dimension of the source and (λ) is the wavelength. Example: If the source is a circle illuminated uniformly and with collimated light, then (s) is the radius of the circle.

Note: Except for scale, the far field diffraction pattern of a source may be observed in the focal plane of a well corrected lens. The far field pattern of a screen illuminated by a point source may be observed in the image plane of the source. Synonym: Fraunhofer diffraction pattern. See also: diffraction; near field diffraction pattern; near field region.

far field radiation pattern—See radiation pattern.

far field region—The region, far from a source, where the diffraction pattern is substantially that observed at infinity. See also: beam divergence angle; diffraction; far field diffraction pattern; near field diffraction pattern; near field region; radiation angle.

ferrule—A mechanical fixture, generally a rigid tube, used to confine the stripped end of a fiber bundle or a fiber. See also: Fiber bundle.

Note: Typically, individual fiber of a bundle are cemented together within a ferrule of a diameter designed to yield a minimum packing fraction. See also: Packing fraction.

Note 2: Nonrigid materials such as shrink tubing may also be used for ferrules for special applications.

fiber—See optical fiber; optical waveguide.

fiber buffer—A material that may be used to protect an individual optical fiber waveguide from physical damage, providing mechanical isolation and/or protection.

Note: Core fabrication techniques vary, some resulting in firm contact between fiber and protective buffering, others resulting in a loose fit, permitting the fiber to slide in the buffer tube. Multiple buffer layers may be used for added fiber protection.

fiber bundle—An assembly of unbuffered optical fibers. A bundle is usually used as a single transmission channel, as opposed to multifiber cables, which contain optically and mechanically isolated fibers, each of which provides a separate channel.

Note 1: Bundles used only to transmit light, as in optical communications, are flexible and are typically unaligned.

Note 2: Bundles used to transmit optical images may be either flexible or rigid, but must contain aligned fibers. See also: aligned bundle; ferrule; fiberoptics; multiwaveguide cable; optical cable; optical fiber; packing fraction.

fiber harness—In equipment interface applications, an assembly of a number of multiple fiber bundles or cables fabricated to facilitate installation into a system.

fiberoptics (FO)—The branch of optical technology concerned with the transmission of radiant power through fibers made of transparent materials, such as glass, fused silica or plastic.

Note 1: Communications applications of fiber optics employ flexible fibers. Either a single discrete fiber or a nonspatially aligned fiber bundle may be used for each information channel. Such fibers are generally referred to in this document as "optical waveguides" to differentiate from fibers employed in noncommunications applications.

Note 2: Various industrial and medical applications employ (typically high-loss) flexible fiber bundles in which individual fibers are specially aligned, permitting optical relay of an image. An example is the *endoscope*.

Note 3: Some specialized industrial applications employ rigid (fused) aligned fiber bundles for image transfer. An example is the fiberoptics *CRT* faceplate used on some high-speed oscilloscopes. See also: optical fiber; optical waveguide.

flux—Obsolete synonym for radiant power.

FO—Abbreviation for fiberoptics.

Fraunhofer diffraction pattern—Synonym for far field diffraction pattern.

Fresnel diffraction pattern—Synonym for near field diffraction pattern.

Fresnel reflection—The reflection of a portion of incident light at a planar interface between two homogenous media having different refractive indices.

Note 1: Fresnel reflection occurs at the air-glass interfaces at entrance and exit ends of an optical waveguide. Resultant transmission losses (on the order of 4% per interface) can be virtually eliminated by use of antireflection coatings or index matching materials.

Note 2: Fresnel reflection depends upon the index difference and the angle of incidence; it is zero at Brewster's angle for one polarization. In optical elements, a thin transparent film is sometimes used to give an additional Fresnel reflection that cancels the original one by intereference. This is called an antireflection coating. See also: antireflection coating; Brewster's angle; index matching materials; reflectance; reflection; reflection loss; refractive index.

fundamental mode—The lowest order mode of a dielectric cylindrical waveguide, in most cases the mode designated (LP_{01}) or (HE_{11}). See also: mode.

fused fiber splice—A splice accomplished by the application of localized heat sufficient to fuse or melt the ends of two lengths of optical fiber, forming, in effect, a continuous, single fiber. See also: mechanical splice; splice.

fused silica—Amorphous silicon dioxide.

Note: Highly refined fused silica formed by a vapor deposition process or by other means is employed in the fabrication of low loss optical waveguides. Dopants may be added via the same process to obtain suitable index variations in the optical waveguide core and cladding regions.

gap loss—An optical power loss caused by a space between source and optical waveguide, between axially aligned waveguides, or between waveguide and detector.

Note: For waveguide-to-waveguide coupling, it is commonly called "longitudinal offset loss." See also: angular misalignment loss; extrinsic junction loss; intrinsic junction loss; lateral offset loss.

gaussian beam—A beam of light whose radial intensity distribution is Gaussian. When such a beam is cylindrical in cross section:

$$I(R) = I(O) \exp\left[- (R/w)^2\right]$$

where (R) is the distance from beam center and (w) is the radius at which the irradiance drops to $(\frac{1}{e})$ of its value on the axis.

gaussian pulse—A pulse that has the shape of a Gaussian distribution. In the time domain, the shape is:

$$f(t) = A \exp [(t / d)^2]$$

where (A) is a constant, and (d) is the pulse halfwidth at the $\left(\frac{1}{e}\right)$ points. A similar expression would hold in the frequency domain with (t) replaced by (v).

geometric optics—The science that treats the propagation of light as rays. Rays are bent at the interface between two dissimilar media, or may be curved in a medium whose refractive index is a function of position. See also: axial ray; meridional ray; optical axis; paraxial ray; physical optics; skew ray.

graded index optical waveguide—A waveguide having a graded index profile. See also; graded index profile; step index optical waveguide.

graded index profile—Any index profile that varies smoothly with radius. Distinguished from a step index profile. See also: dispersion; index profile; mode volume; multimode optical waveguide; normalized frequency; optical waveguide; parabolic profile; profile dispersion; profile parameter; refractive index; step index profile; power-law index profile

group degenerate modes—Modes that have the same group velocity. See also: mode; phase degenerate modes.

group index—Denoted N: Velocity of light in vacuum divided by the group velocity in a medium of index (n). It is related thus to the refractive index:

$$N = m - \lambda \frac{dn}{d\lambda}$$

See also: group velocity; material dispersion parameter.

group velocity—Velocity of the signal modulating a propagating electromagnetic wave. It is given by (c/N) where (c) is the velocity of light in vacuum and (N) is the group index. See also: differential modal delay; group degenerate modes; group index; phase velocity.

guided mode—Synonym for bound mode.

(HE) mode—See fundamental mode.

hybrid mode—A mode possessing components of both electric

and magnetic field vectors in the direction of propagation.
Note: Such modes correspond to skew (non-meridional) rays.
See also: mode; skew ray; transverse electric mode; transverse magnetic mode.

ILD—Abbreviation for injection laser diode.

impulse response—The function h(t) describing the response of an initially relaxed system to an impulse (Dirac-delta) function applied at time t = 0.
Note. The impulse response may be obtained by deconvolving the input waveform from the output waveform, or as the inverse Fourier transform of the transfer function. See also: transfer function.

incoherent—Characterized by a degree of coherence significantly less than 0.88. See also: coherent; degree of coherence.

index matching materials—Transparent materials of proper refractive index used to reduce Fresnel reflections at an optical interface. See also: Fresnel reflection; mechanical splice; refractive index.

index of refraction—Synonym for refractive index.

index profile—In an optical waveguide, the refractive index as a function of radius. See also: graded index profile; parabolic profile; power-law index profile; profile dispersion; profile dispersion parameter; profile parameter; step index profile.

infrared (IR)—The region of the electromagnetic spectrum between the long-wavelength extreme of the visible spectrum (about $0.7\ \mu$m) and the shortest microwaves (about 1 mm). See also: light; ultraviolet (UV).

injection fiber—Synonym for launching fiber.

injection laser diode (ILD)—A laser employing as the active medium a forward-biased semiconductor diode. Synonym: diode laser. See also: active laser medium; chirping; laser; superradiance.

insertion loss—The total optical power loss in a transmission system caused by the insertion of an optical component such as a connector, splice, or coupler.

integrated optical circuit (IOC)—A monolithic optical circuit, composed of both active and passive miniaturized components, employing planar waveguides for coupling to optoelectronic devices and providing signal processing functions such as modulation, multiplexing and switching.

intensity—The square of the electric field amplitude of a light wave. Intensity is proportional to irradiance and may be used in place of the term "irradiance" when ony relative values are important. See also: irradiance; radiant intensity; radiometry.

interference—In optics, the interaction of two or more beams of coherent or partially coherent light usually derived from a single source. See also: coherent; degree of coherence; diffraction.

intermodal distortion—Synonym for multimode distortion.

intramodal distortion—That distortion resulting from dispersion within individual propagating modes. It is the only distortion occurring in single mode waveguides. See also: dispersion; distortion.

intrinsic junction loss—The total loss resulting from joining two identical optical waveguides. Note. Factors influencing this loss include spacing loss, alignment of the waveguides, Fresnel reflection loss, end finish, etc. See also: angular misalignment loss; gap loss; extrinsic junction loss; lateral offset loss.

IOC—Abbreviation for integrated optical circuit.

ion exchange technique—A method of fabricating a graded index optical waveguide by exchanging ions through the core-cladding interface. See also: chemical vapor deposition (CVD) technique; double crucible technique; graded index profile; rod-in-tube technique; soot technique; vapor phase axial deposition technique.

irradiance—Radiant power incident per unit area upon a surface, expressed in watts per square meter. "Power density" is colloquially used in a synonym. See also: beamwidth: radiometry.

Lambertian radiator—See Lambert's cosine law.

Lambertian reflector—See Lambert's cosine law.

Lambertian source—See Lambert's cosine law.

Lambert's cosine law—The statement that the radiance of certain idealized surfaces, known as Lambertian radiators, Lambertian sources, or Lambertian reflectors, is independent of the angle from which the surface is viewed.

Note: The radiant intensity of such a surface is maximum normal to the surface and decreases in proportion to the cosine of the angle from the normal. Synonym: cosine emission law.

laser—A device that produces optical radiation using a population inversion to provide Light Amplification by Stimulated Emission of Radiation and (generally) an optical resonant cavity to provide positive feedback. *Laser radiation may be highly coherent* either temporarily or spatially, or both. See also: active laser medium: injection laser diode; optical cavity.

laser medium—Synonym for active laser medium.

lasing—Process of emitting laser light.

lasing threshold—The lowest excitation level at which a laser's output is dominated by stimulated emission rather than spontaneous emission.

lateral offset loss—A power loss caused by transverse or lateral deviation from optimum alignment of source to optical waveguide, waveguide to waveguide, or waveguide to detector. Synonym: Transverse Offset Loss. See also: angular misalignment loss; extrinsic junction loss; gap loss; intrinsic junction loss.

launch angle—The angle between the light input propagation vector and the optical axis of an optical fiber or fiber bundle.

launching fiber—A fiber or fibers used in conjunction with a source to excite the modes of a fiber in a particular fashion. Note: Launching Fibers are most often used in test systems to improve the precision of measurements. Synonym: injection fiber. See also: mode; pigtail.

launch numerical aperture (LNA)—The numerical aperture of an optical system used to coupled (launch) power into an optical waveguide. Note 1: (LNA) may differ from the state (NA) of a final focusing element if, for example, that element is underfilled or the focus is other than that for which the element is specified. Note 2: (LNA) is one of the parameters that determine the initial distribution of power among the modes of an optical waveguide. See also: acceptance angle.

leaky modes—In an optical waveguide, those modes that are weakly bound to the core of the waveguide and have comparatively high loss as a result of tunneling. See also: bound mode; cladding mode; mode; unbound mode.

LED—Abbreviation for light emitting diode.

light—In a strict sense, the region of the electromagnetic spectrum that can be perceived by human vision, designated the visible spectrum and nominally covering the wavelength range of 0.4 μm to 0.7 μm.

In the laser and optical communication fields, custom and practice have extended usage of the term to include the much broader portion of the electromagnetic spectrum that can be handled by the basic optical techniques used for the visible spectrum. This region has not been clearly defined but, as employed by most workers in the field, may be considered to extend from the near-ultraviolet region or approximately 0.3 μm, through the visible region, and into the mid-infrared region of 3.0 μm to 30 μm. See also: infrared (IR); optical spectrum; ultraviolet (UV).

light current—See photocurrent.

light emitting diode (LED)—A (pn) junction for semiconductor device that emits *incoherent* optical radiation when biased in the forward direction.

lightguide—Synonym for optical waveguide.

light ray—The path of a given point on a wavefront. The direction of a light ray is generally normal to the wavefront. See also: geometric optics.

linear optical element or system—One in which the radiant power output is proportional to the radiant power input, and no new optical wavelengths or modulation frequencies are generated.

Note 1: The proportionality constant can vary with source wavelength and modulation frequency.

Note 2: A linear element can be described in terms of a transfer function and an impulse response function.

linearly polarized (LP) mode—A mode for which the field components in the direction of propagation are small compared to components perpendicular to that direction.

Note: Each (LP) mode consists of several phase degenerate modes. The (LP) description (which is an approximation for weakly guiding waveguides) becomes more accurate as the difference between the maximum and minimum values of the refractive index becomes a smaller fraction (typically < 2%) of the mean index value across the profile. See also: mode; phase degenerate mode; weakly guiding optical waveguide.

line source—1. In the spectral sense, an optical source that emits one or more spectrally narrow lines as opposed to a continuous spectrum. See also: monochromatic. 2. In the geometric sense, an optical source whose active (emitting) area forms a spatially narrow line.

line spectrum—An emission or absorption spectrum consisting of one or more narrow spectral lines, as opposed to a continuous spectrum. See also: monochromatic; spectral line; spectral linewidth.

linewidth—See spectral linewidth.

LNA—Abbreviation for launch numerical aperture.

longitudinal offset loss—See gap loss.

loss—See absorption; angular misalignment loss; attenuation; backscattering; extrinsic junction loss; gap loss; insertion loss; intrinsic junction loss; lateral offset loss; macrobend loss; material scattering; microbend loss; nonlinear sacttering; Rayleigh scattering; reflection loss; transmission loss; waveguide scattering.

lp mode—Abbreviation for linearly polarized mode.

LP$_{01}$ mode—See fundamental mode.

macrobending—In an optical waveguide, all macroscopic axiaι deviations of the waveguide from a straight line, as opposed to microbending. See also Microbending.

macrobend loss—In an optical waveguide, that portion of the total loss attributable to macrobending. Synonym: curvature loss. See also: microbend loss.

magneto-optic—Pertaining to a change in a material's refractive index under the influence of a magnetic field. Magneto-optic materials generally are used to rotate the plane of polarization.

material absorption—See absorption.

material dispersion—That dispersion attributable to the wavelength dependence of the refractive index of material used to form the waveguide. Material dispersion is characterized by a parameter (M) which is defined below. See also: dispersion; distortion; profile dispersion; profile dispersion parameter; waveguide dispersion.

material dispersion parameter (M)—

$$n(\lambda) = -\frac{1}{c}\left(\frac{dn}{d\lambda}\right) = \frac{\lambda}{c}\ (d^2{}_n/d\lambda^2)$$

where (n) is the refractive index,

(N) is the material group index: $N = n - \lambda\left(\dfrac{dn}{d\lambda}\right)$

(λ) is the wavelength and,

(c) is the velocity of light in vacuum.

Note 1: For many present optical waveguide materials, (M) is zero at a specific wavelength λ_o, usually found in the 1.2 to 1.5 μm range. The sign convention is such that (M) is positive for wavelengths shorter than (λ_o) and negative for wavelengths longer than (λ_o).

Note 2: Pulse broadening caused by material dispersion in an optical fiber is given by (M) times spectral linewidth ($\Delta\lambda$), except at ($\lambda \cong \lambda_o$). (See Note 1). See also: Group Index.

material scattering—In an optical waveguide, that part of the total scattering attributable to the properties of the bulk materials used for waveguide fabrication.

Note: Material scattering may be either intrinsic scattering resulting from frozen-in inhomogeneities, or extrinsic scattering resulting from impurities. See also: Rayleigh scattering; scattering; waveguide scattering.

mechanical splice—An optical waveguide splice accomplished by external fixtures or materials, rather than by thermal fusion. Index matching material may be applied between the two fiber ends. See also: fused fiber splice; index matching material.

meridional ray—A ray that crosses through the optical axis of an optical waveguide (in contrast with a skew ray). See also: axial ray; geometric optics; numerical aperture; optical axis; paraxial ray; skew ray.

microbend loss—In an optical waveguide, sharp curvatures involving local axial displacements of a few micrometers and spatial wavelengths of a few millimeters. Such bends may result from waveguide coating, cabling, packaging, installation, etc. See also: macrobending.

misalignment loss—See angular misalignment loss; gap loss; laterial offset loss.

mode—In any cavity or transmission line, one of the allowed electromagnetic field distributions. The field pattern of a given mode depends on the wavelength, refractive index, and cavity or waveguide geometry. See also: bound mode; cladding mode; degenerate waveguide modes; differential modal attenuation; differential modal delay; equilibrium mode distribution; equilibrium mode simulator; fundamental mode; group degenerate modes; hybrid mode; leaky modes; linearly polarized mode; mode volume; multimode distortion; multimode laser; multimode optical waveguide; phase degenerate modes; single mode optical waveguide; transverse electric mode; transverse magnetic mode; unbound mode.

mode coupling—In an optical waveguide, the exchange of power among modes.

Note: Mode coupling reaches equilibrium after propagation over a finite distance that is designated the equilibrium length. See also: equilibrium length; equilibrium mode distribution; mode; mode scrambler.

mode dispersion—Often erroneously used as a synonym for mode distortion.

mode (or modal) distortion—Synonym for multimode distortion.

mode filter—A device used to attenuate certain modes.

mode mixer—Synonym for mode scrambler.

mode scrambler—A device for inducing mode coupling. Synonym: mode mixer. See also: mode coupling.

mode volume—The number of propagating modes that an optical waveguide will support; for (V>5), approximately given by $\left(\dfrac{V^2}{2}\right)$ and $\left(\dfrac{V^2}{2}\right)\left(\dfrac{g}{g+2}\right)$ respectively, for step index and power-law profile waveguides, where (g) is the profile parameter. See also: mode; normalized frequency; power-law index profile; step index profile.

modulation—A controlled variation with time of any property of wave for the purpose of transferring information.

monochromatic—An idealized concept referring to a single frequency or wavelength. In practice, radiation is never perfectly monochromatic but, at best, displays a narrow band of wavelengths. See also: coherent; line sources; spectral linewidth.

monomode optical waveguide—Synonym for single mode optical waveguide.

multifiber cable—An optical cable that contains two or more optical waveguides, each of which provides a separate information channel. See also: fiber bundle; optical cable assembly.

multifiber connector—An optical connector designed to mate two multifiber cables, providing simultaneous optical alignment of all individual waveguides.

Note: Optical coupling between aligned waveguides may be achieved by various techniques including proximity butting (with or without index matching materials), and the use of relay optics (solid or liquid lenses).

multilayer dielectric filter—An optical filter consisting of a sequence of thin layers of transparent material with controlled thicknesses and refractive indices. See also: dichroic filter.

multimode distortion—In an optical waveguide, that distortion resulting from the superposition of modes have differential modal delays.

Note: The term "multimode dispersion" is often used as a synonym; such usage, however, is erroneous since the mechanism is not dispersive in nature. Synonym: intermodal distortion; mode (or modal) distortion. See also: differential modal delay; dispersion; distortion; mode; multimode optical waveguide.

multimode group delay—Synonym for differential modal delay.

multimode laser—A laser that produces simultaneous emission at two or more discrete wavelengths and/or in two or more transverse modes. See also: laser; mode.

multimode optical waveguide—An optical waveguide that will allow more than one bound mode to propagate.

Note: May be either a graded index or step index waveguide. See also: bound mode; index profile; mode; mode volume; multimode distortion; normalized frequency; power-law index profile; single mode optical waveguide; step index optical waveguide.

multiplexing—See wavelength division multiplexing (WDM).

NA—Abbreviation for numerical aperture.

near field diffraction pattern—The diffraction pattern observed close to a source of aperture, as distinguished from far field diffraction pattern. Synonym: Fresnel diffraction pattern. See also: diffraction; far field region; far field diffraction pattern.

near field radiation pattern—See radiation pattern.

near field region—The region close to a source, or aperture. The Diffraction pattern in this region typically differs significantly from that observed at infinity and varies with distance from the source. See also: far field diffraction pattern; far field region.

nephroscope—A flexible cable containing a fiber-optic lens and light system used in kidney stone disintegration.

noise equivalent power (NEP)—At a given modulation frequency and for a given bandwidth, the radiant power that produces a signal-to-noise ratio of 1 at the output of a given detector. In this sense, the minimum detectable power at the given frequency and for the given bandwidth.

Note: Some manufacturers and authors define NEP as the minimum detectable power per unit bandwidth; when defined in this way, NEP has the units of watts/(hertz) $1/2 = \sqrt{\dfrac{w}{f}}$.

Therefore, use of the term NEP for a quantity whose units are watts/(hertz) 1/2 is a misnomer, because the units of power are watts. See also: D*: detectivity.

nonequilibrium mode distribution—That distribution of modes prevailing in a length of waveguide shorter than the equilibrium length. See also: equilibrium mode distribution.

nonlinear scattering—Direct conversion of a single photon from one wavelength to one or more other wavelengths. In an optical waveguide, nonlinear scattering is usually not important below the threshold irradiance to stimulated nonlinear scattering.

Note: Examples are Raman and Brillouin scattering. See also: photon.

normalized frequency—A dimensionless quantity (denoted by V), mathematically given by:

$$V = \frac{2\pi a}{\lambda} \sqrt{N_1^2 - N_2^2}$$

where a is waveguide core radius, λ is wavelength in vacuum, and N_1 and N_2 are refractive indices at the waveguide axis and in the cladding, respectively. The number of bound modes in a waveguide is approximately proportional to V^2; in a step index waveguide, the number of modes is $V^2/2$; in a parabolic guide $V^2/4$. Synonym: V number. See also: bound mode; mode volume; parabolic profile; single mode optical waveguide.

numerical aperture (NA)—NA = N sin θ where θ is, at a specified point, half the vertex angle of the largest cone of meridional rays that can enter or leave an optical element or system, and N is the refractive index of the homogeneous isotropic space that contains the specified point. The specified point is usually an object or image point.

Note: The term "numerical aperture" is often used, imprecisely, to describe an optical waveguide. The precise terms "acceptance angle" and "radiation angle" are preferred. See also: acceptance angle; equilibrium radiation angle; equilibrium radiation pattern; launch numerical aperture; meridional ray; radiation angle; radiation pattern.

optical axis—In a cylindrically symmetric waveguide, the optical axis is the axis of geometric symmetry of the core. Distinguished from "optic axis."

Note 1: In rare cases, an optical system will contain one or more elements (usually mirrors) whose axes are tilted with respect to one another. In such cases the optical axis is ill defined but could be taken to follow the line segments that join the centers of the aperture stops of the tilted components.

Note 2: The optical axis of a waveguide is the axis of the waveguide and therefore need not be a straight line. See paraxial ray; skew ray.

optical blank—A casting consisting of an optical material molded into the desired geometry for grinding, polishing, or (in the case of optical wavelengths) drawing to the final optical/mechanical specifications. See also: preform.

optical cable—A fiber, multiple fibers, or fiber bundle in a structure fabricated to meet optical, mechanical, and environmental specifications. Synonym: optical fiber cable.

optical cable assembly—A cable that is connector terminated. Generally, a cable that has been terminated by a manufacturer and is ready for installation.

optical cavity—A region bounded by two or more reflecting surfaces, referred to as mirrors, end mirrors, or cavity mirrors, whose elements are aligned to provide multiple reflections. The resonator in a laser is an optical cavity. Synonym: resonant cavity. See also: active laser medium; laser.

optical conductor—Synonym for optical waveguide (not recommended).

optical connector—See optical waveguide connector.

optical coupler—See optical waveguide coupler.

optical data bus—A data bus using optical waveguides and optical waveguide components.

optical density—A measure of the transmittance of an optical element expressed by: $\log_{10}\left(\dfrac{2}{T}\right)$ OR $-\log_{10} T$ where T is transmittance. The analogous term $\log_{10}\left(\dfrac{1}{R}\right)$ is called reflection density.

Note: The higher the optical density, the lower the transmittance. Optical density times -10 is equal to transmittance loss expressed in decimals; for example, the transmittance loss of -3 dB is equal to the optical density of 0.3. See also: transmittance loss; transmittance.

optical detector—A transducer that generates an electrical signal that is a function of irradiance.

optical fiber—Any filament or fiber, made of dielectric materials, that guides light whether or not it is used to transmit signals. See also: fiber bundle; fiberoptics; optical waveguide; signal.

optical-fiber cable—Synonym for optical cable.

optical-fiber waveguide—Synonym for optical waveguide.

optical filter—An element that transmits a range of wavelengths and blocks adjacent wavelengths. See also: dichroic filter; multilayer dielectric filter.

optical link—Any optical transmission channel designed to interconnect two end terminals (constituting a circuit in communications terminology) or to be connected in series (as part of a circuit) with other channels.

Note: Sometimes terminal hardware (e.g., transmitter/receiver modules) is included in the definition.

optical path length—In a medium of constant refractive index (n), the product of the geometrical distance and the refractive index. If n is a function of position, then optical path length $= \int$ nds. where ds is an element of length along the path. This expression simplifies to n if the medium has a constant refractive index. See also: optical thickness.

optical power—Colloquial synonym for radiant power.

optical repeater—In an optical waveguide communication system, an optoelectronic device or module that receives an input optical signal, converts it into an electrical signal, amplifies this signal (or, in the case of a digital signal, reshapes, retimes, or otherwise reconstructs the signal) and reconverts it into an optical signal for retransmission.

optical spectrum—Bandwidth limits have not been rigidly defined, but, as employed by most workers in the field, the region of the electromagnetic spectrum within the wavelength region extending from the vacuum ultraviolet at 40 nm to the far infrared at 1 mm.

optical thickness—Applied to thin films, the physical thickness times the refractive index. See also: optical path length.

Optical waveguide—

1. Any structure capable of guiding optical power.

2. In optical communications, generally a fiber designed to transmit optics' signals. Synonym: lightguide; optical conductor (deprecated); optical fiber waveguide; optical waveguide fiber. See also: cladding; core; fiber bundle; fiberoptics; multimode optical waveguide; optical fiber; single mode optical waveguide; tapered fiber optical waveguide.

optical waveguide connector—A device whose purpose is to transfer optical power between two optical waveguides or bundles, and that is designed to be connected and disconnected repeatedly. See also: multifiber connector.

optical waveguide coupler—

1. A device whose purpose is to distribute optical power among three or more ports. See also: star coupler; tee coupler.

2. A device whose purpose is to couple optical power between a waveguide and a source or detector.

optical waveguide fiber—Synonym for optical waveguide.

optical waveguide splice—A permanent joint whose purpose is to couple optical power between two waveguides. See also: fusion splice; mechanical splice.

optically active material—A material that has the ability to rotate the plane of polarization of light that propagates through it. Note: An optically active material exhibits different refractive indices for left and right circular polarizations (circular birefringence).

optic axis—In an anisotropic medium, a direction of propagation in which orthogonal polarizations have the same phase velocity. Distinguished from "optical axis."

optoelectronic—Pertaining to a device that responds to optical power, emits or modifies optical radiation, or utilizes optical radiation for its internal operation. Any device that functions as an electrical-to-optical or optical-to-electrical transducer. Note 1: Photodiodes, LEDs, injection lasers and integrated optical elements are examples of optoelectronic devices commonly used in optical waveguide communications. Note 2: "Electro-optical" is often erroneously used as a synonym.

output angle—Synonym for radiation angle.

packing fraction—In a fiber bundle, the ratio of the aggregate fiber core area to the total cross-sectional area (usually within the ferrule) including cladding and interstitial areas. See also: ferrule; fiber bundle.

parabolic profile—A power-law index profile with the profile parameter (g) equal to 2. Synonym: quadratic profile. See also: graded index profile; index profile, multimode optical waveguide; power-law index profile; profile parameter.

paraxial ray—A ray that is close to and nearly parallel with the optical axis. Note: For purposes of computation, the angle between the ray and the optical axis is small enough for (sin θ) or (tan θ) to be replaced by [θ (radians)]. See also: axial ray; geometric optics; meridonal ray; skew ray.

peak wavelength—The wavelength at which the radiant intensity of a source is a maximum. Peak wavelength is typically expressed in nanometers.

phase coherence—See Coherent.

phase degenerate modes—Modes that have the same phase velocity. See also: group degenerate modes; linearly polarized mode; mode.

phase velocity—Velocity of the signal modulating a propagating electromagnetic wave. It is given by c/n where c is the velocity of light in vacuum and n is the refractive index. See also: coherence time; group velocity; phase degenerate modes.

photoconductivity—The conductivity increase exhibited by some nonmetallic materials, resulting from the free carriers generated when photon energy is absorbed in electronic transitions. The rate at which free carriers are generated, the mobility of the carriers, and the length of time they persist in conducting states (their lifetime) determine the amount of conductivity change. See also: photoelectric effect.

photocurrent—The current that flows through a photosensitive device (such as a photodiode) as the direct result of exposure to radiant energy internal gain mechanisms, such as in an avalanche photodiode, may enhance or increase the electron flow but are distinct mechanisms. See also: dark current.

photodiode—A diode having a current-vs-voltage characteristics that is dependent on the level of optical power incident on the device. Photodiodes are used for the detection of optical power and for the conversion of optical power to electrical power. See also: avalanche photodiode (APD); PIN photodiode.

photoelectric effect—External photoelectric effect: The emission of electrons from the irradiated surface of a material into a vacuum or a gas. Synonym: photoemissive effect.

photoemissive effect—Synonym for photoelectric effect.

photon—A quantum of electromagnetic energy. The energy of a photon is hv where h is Planck's constant and v is the optical frequency. See also: nonlinear scattering; Planck's constant.

photovoltaic effect—The production of a voltage across a pn junction resulting from the absorption of photon energy. The potential is caused by the internal drift of hole-electron pairs; hence the phenomenon leads to direct conversion of a part of the absorbed energy into a usable voltage.

physical optics—The branch of optics in which light propagation is treated as a wave phenomenon rather than a ray phenomenon, as in geometric optics. See also: geometric optics.

pigtail—A short length of optical fiber used to couple power between an optoelectronic component and the transmission fiber. See also: launching fiber.

PIN photodiode—A diode with a large intrinsic (very lightly doped) region sandwiched between (p) and (n) doped semiconducting regions. Photons absorbed in this region create electron-hole pairs that are then separated by an electric field, thus generating an electric current in a load circuit. See also: avalanche photodiode (APD); photodiode.

Planck's constant—(Denoted h): $h = 6.626 \times 10^{-34}$ joule second.

plane wave—A wave whose surfaces of constant phase are parallel planes normal to the direction of propagation and infinite in extent.

Pockel's effect—See electro-optic effect.

power—See irradiance; power efficiency; radiant power.

power density—Colloquial synonym for irradiance.

power-law index profile—A class of graded index profiles characterized by the following equations:

$$n(r) = N_1 \left[1 - 2\Delta \left(\frac{r}{a} \right)^g \right]^{\frac{1}{2}} \quad r \leq a$$

$$n(r) = N_2 = N (1 - 2\Delta)^{\frac{1}{2}} \quad r \geq a$$

$$\text{where} \quad \Delta = \frac{N_1^2 - N_2^2}{2 N_1^2}$$

where n(r) is the refractive index as a function of radius, n_1 is the refractive index on axis, n_2 is the refractive index of the cladding, a is the core radius, and g is a parameter that defines the shape of the profile.

Note: For this class of profiles, multimode distortion is smallest when g takes a particular value depending on the material used. For most materials, this optimum value falls around 2. When g is very large, the profile becomes a step index profile. See also: graded index profile; index profile; mode volume; profile parameter; step index profile.

preform—A glass structure from which an optical fiber waveguide may be drawn. See also: boule; chemical vapor deposition technique; ion exchange technique; optical blank; rod-in-tube technique; soot technique; vapor phase axial deposition technique.

profile—See graded index profile; index profile; parabolic profile; power-law index profile; step index profile.

profile dispersion—In an optical waveguide, that dispersion attributable to the variation of refractive index profile with wavelength. See also: dispersion; material dispersion; waveguide dispersion.

profile dispersion parameter (P)—

$$P(\lambda) = \frac{n_o}{N_o} \frac{\lambda}{\Delta} \frac{d\Delta}{d\lambda}$$

where (n_o), (N_o) are respectively the refractive and group indices at the core center, and $n_o \sqrt{1-2\Delta}$ is the phase index at the core edge or cladding. The expression is uniquely specified for power-law index profiles. Sometimes it is defined with the factor -2 in the numerator.

profile parameter—The shape-defining parameter, (g), for a power-law index profile. See also: power-law index profile.

pulse distortion—See distortion.

pulse duration—The time between a specified reference point on the first transition of a pulse waveform and a similarly specified point on the last transition. The time between the 10%, 50%, or $1/e$ power points are commonly used, as is the rms pulse duration; therefore, the measurement level must be stated in quantitative use of the term. See also: root mean square pulse duration.

pulse length—Often erroneously used as a synonym for pulse duration.

pulse width—Often erroneously used as a synonym for pulse duration.

Q switch—A device that prohibits oscillation of a laser until the energy stored in the active medium increases to a desired level. Note: In a pulsed laser, a Q switch increases peak power by shortening pulse durations; the device provides shorter and more powerful pulses than would be possible by direct electrical or optical switching. Q switching is most effective with glass or crystal cavity lasers and some varieties of gas lasers.

quadratic profile—Synonym for parabolic profile.

quantum efficiency—A dimensionless measure of the efficiency of conversion or utilization of optical energy, being the average number of charged carriers produced for each incident photon.

quantum limited operation—Operation wherein the minimum detectable signal is limited by quantum noise.

quantum noise—Any noise attributable to the discrete nature of electromagnetic radiation. Examples include shot noise, photon noise, and recombination noise. See also: shot noise.

radiance — Radiant power, in a given direction, per unit solid angle per unit of projected area of the source, as viewed from that given direction. Radiance is expressed watts per steradian per square meter. See also: brightness; conservation of radiance; radiometry.

radiance theorem — Synonym for conservation of radiance.

radiant emittance — Radiant power emitted into a full sphere (4 π steradians) by a unit area of a source; express in watts per square meter. Synonym: radiant exitance. See also: radiometry.

radiant energy — Energy that is transferred via electromagnetic waves, i.e., the time integral of radiant power; expressed in joules. See also: radiometry.

radiant exitance — Synonym for radiant emittance.

radiant flux — Synonym for radiant power (obsolete).

radiant incidence — See irradiance.

radiant intensity — Applied to a point source only, the time rate of transfer of radiant energy per unit solid angle, expressed in watts per steradian. See also: intensity, radiometry.

radiant power — The time rate of low of radiant energy, expressed in watts. When exclusive concern with the optical spectrum is assumed, the prefix is often dropped and the term "power" is used. Colloquial synonym: flux; optical power; power; radiant flux.

radiation angle — Half the vertex angle of that cone within which can be found a specified fraction of the total radiated power at any distance in the far field. Synonym: output angle. See also: acceptance angle; equilibrium radiation angle; equilibrium radiation pattern; far field region; numerical aperture.

radiation pattern — The output radiation of an optical waveguide, specified as a function of angle or distance from the waveguide axis.

Note 1: Far field radiation pattern is specified as a function of angle. Near field radiation pattern is specified as a function of distance from the waveguide axis.

Note 2: Radiation pattern is a function of the length of waveguide measure, the manner in which the waveguide is excited, and the wavelength. See also: acceptance angle; equilibrium radiation angle; equilibrium radiation pattern, far field region; near field region; numerical aperture.

radiative mode—Synonym for unbound mode.

radiometry—The science of radiation measurement. The basic quantities of radiometry are listed below:

Radiometric terms

Term name	Symbol	Quantity	Unit
Radiant Energy	Q	Energy	Joule (J)
Radiant Power Synonym: Radiant Flux	ϕ	Power	Watt (W)
Irradiance	E	Power incident per unit area (irrespective of angle)	$W \cdot m^{-2}$
Spectral Irradiance	E_λ	Irradiance per unit wavelength interval at a given wavelength	$W \cdot m^{-2} \cdot nm^{-1}$
Radiant Emittance Synonym: Radiant Exitance	W	Power emitted (into a full sphere) per unit area	$W \cdot m^{-2}$
Radiant Intensity	I	Power per unit solid angle	$W \cdot steradian^{-1}$
Radiance	L	Power per unit solid angle per unit projected area	$W \cdot steradian^{-1} \cdot m^{-2}$
Spectral Radiance	L_λ	Radiance per unit wavelength interval at a given wavelength	$W \cdot steradian^{-1} \cdot m^{-2} \cdot nm^{-1}$

ray—See light ray.

Rayleigh scattering—Scattering by submicroscopic inhomogeneities (fluctuations in material density or composition) in refractive index. A feature of Rayleigh scattering is that the scattered field is inversely proportional to the fourth power of the wavelength. See also: material scattering; waveguide scattering; scattering.

reflectance—The ratio of reflected power to incident power. Note: In optics, frequently expressed as optical density or as a percent; in communications applications; generally expressed in dB. Reflectance may be defined as specular or diffuse, dependant on the nature of the reflecting surface. Formerly: "reflection."

See also: reflection; reflection loss; reflectivity.

reflection—The abrupt change in direction of a light beam at an interface between two dissimilar media so that the light beam returns into the medium from which it originated. Reflection from a smooth surface is termed specular, whereas reflection from a rough surface is termed diffuse. See also: critical angle; reflectance; reflectivity; total internal reflection.

reflection loss—Total loss from reflections at the junction between two optical components. See also: antireflection coating; Fresnel reflection; reflectance; reflectivity.

reflectivity—The reflectance of the surface of a material which is so thick that the reflectance does not change with increasing thickness; the intrinsic reflectance of the surface, irrespective of other parameters such as the reflectance of the rear surface. No longer in common usage.

refraction—The bending of a beam of light in transmission through an interface between two dissimilar media or in a medium whose refractive index is a continuous function of position (graded index medium). See also: angle of deviation; refractive index.

refractive index—The radio of the velocity of light in vacuum to the plane velocity in a medium, a function of wavelength; denoted by n. May also be defined as the square root of relative permitivity. Synonym: index of refraction. See also: cladding; core; critical angle; dispersion; Fresnel reflection; fused silica; graded index optical waveguide; linearly polarized mode; material dispersion; mode; normalized frequency; numerical aperture; optical path length; power-law index profile; profile dispersion; scattering; step index optical waveguide; weakly guiding optical waveguide.

regenerative repeater—A repeater that is designed for digital transmission (may use a form of positive feedback). See also: optical repeater.

repeater—See optical repeater.

responsivity—The ratio of an optical detector's electrical output to its' optical input, the precise definition depending on the detector type; generally expressed in amps per watt or volts per watt of incident radiant power.

Note: "Sensitivity" is often incorrectly used as a synonym.

rms pulse duration—See root mean square (rms) pulse duration.

rod-in-tube technique—A method of fabricating an optical waveguide by placing a rod in a tube and drawing the rod and the tube to form a fiber.

root-mean-square (rms) pulse duration—A measure of the duration of a pulse waveform. Specifically:

$$\theta_{rms} = [1 \ \frac{1}{A} \int (t-T)^2 \ S(t) \ dt]^{1/2}$$

where S(t) is the amplitude of the pulse as a function of time, A is given by A = S(t) dt and T is the central time, given by T = ∫ t S(t) dt

Note: The ∫ dt implies integration over all time.

scattering—The deflection of light from the path it would follow if the refractive index were uniform or gradually graded.
Note 1: Scattering is caused primarily by microscopic or submicroscopic fluctuations in refractive index.
Note 2: In a waveguide, scattering results in mode coupling, one effect of which is to transfer power from bound modes into leaky and radiative modes, and into backward traveling modes. An observable effect is light emerging from the sides of the guide, resulting in signal attenuation. See also: leaky modes; material scattering; mode; nonlinear scattering; Rayleigh scattering; refractive index; unbound mode; waveguide scattering.

self focusing fiber—Synonym for graded index optical waveguide.

semiconductor laser—Synonym for injection laser.

sensitivity—Imprecise synonym for responsivity.

shot noise—Noise caused by current fluctuations due to the discrete nature of charge carriers and random and/or unpredictable emission of charged particles from an emitter.
Note: There is often a (minor) inconsistency in notation when referring to shot noise in an optical system; many authors refer to shot noise loosely when speaking of the mean square shot noise current (amp^2) rather than noise power (watts).

signal—The information of intelligence that is transferred over a comunications system by electrical or optical means.

signal mode optical waveguide—An optical waveguide only one bound mode (frequency) can propagate at the wavelength of interest. In step index guides, this occurs when the normalized frequency (v) is less than 2.405. For power-law profiles, single mode operation occurs for normalized frequency (v) less than approximately $2.405 \sqrt{\dfrac{g+2}{g}}$ where g is the profile parameter.

Synonym: monomode optical waveguide. See also: bound mode; index profile; mode; multimode optical waveguide; normalized frequency; power-law index profile; profile parameters; step index optical waveguide.

skew ray—A ray that does not intersect the optical axis of a system (in contrast with a meridonal ray).

soot technique—A method of fabricating an optical waveguide perform by forming small glass particles (soot) and depositing the particles on the surface of a cylinder. See also: chemical vapor deposition technique; double crucible technique; ion exchange technique; rod-in-tube technique; vapor phase axial deposition technique.

source efficiency—The ratio of emitted optical power of a source to the input electrical power.

spatial coherence—See coherent.

spatially aligned bundle—See aligned bundle.

spatially coherent radiation— See coherent.

specific detectivity—Synonym for D*.

spectral irradiance—Irradiance per unit wavelength interval at a given wavelength, expressed in watts per square meter per micrometer.

spectral line—A narrow range of emitted or absorbed wavelengths. See also: line source; line spectrum; monochromatic.

spectral linewidth—A measure of the purity of a spectral line occurring in a line spectrum.

Note 1: One method of specifying the spectral linewidth is the full width at half maximum (FWHM), specifically the difference between the wavelengths at which the spectral emittance or absorption drops to one half of its maximum value. This method may be difficult to apply when the line has a complex shape.

Note 2: Another method of specifying the spectral linewidth is the rms width, given by:

$$\Delta\lambda = [\ \frac{1}{A}\ \int\ (\lambda-\lambda_o)^2\ S(\lambda)\ d\lambda]^{\frac{1}{2}}$$

where $S(\lambda)$ is a suitable radiometric quantity, A is given by:

$$A = \int S(\lambda)d\lambda$$

and λ_o is the central wavelength given by:

$$\lambda_o = \frac{1}{A}\ \int\ \lambda S(\lambda)\ d\lambda$$

Note 3: The relative spectral linewidth $\frac{\Delta\lambda}{\lambda}$ is frequently used.

spectral radiance—Radiance per unit wavelength interval at a given wavelength, expressed in watts per steradian (solid angle) per square centimeter per micrometer. See also: radiance; radiometry.

spectral responsivity—Responsivity per unit wavelength interval at a given wavelength.

spectral window—A wavelength region at which relatively minimal attentuation of an optical signal is experienced. Synonym: transmission window.

spectrum—Optical spectrum.

specular reflection—See reflection.

splice—See optical waveguide splice.

splice loss—See insertion loss.

spontaneous emission—Radiation emitted when the internal energy of a quantum mechanical system drops from an excited level to a lower level without regard to the simultaneous presence of similar radiation.

Note: Examples of spontaneous emission include: 1) radiation from an LED, and 2) radiation from an injection laser below the lasing threshold. See also: injection laser diode; light-emitting diode; stimulated emission; superradiance.

star coupler—A passive coupler whose purpose is to distribute optical power from one port among all ports (reflection star) or a set of all other ports (transmission star). See also: tee coupler.

steady state condition—Synonym for equilibrium mode distribution.

step index optical waveguide—An optical waveguide having a step index profile.

step index profile—An index profile characterized by a uniform refractive index within the core and a sharp step decrease in refractive index at the core-cladding interface, or at various layers of the core.

Note: This corresponds to a power-law profile with profile parameter, g, approaching infinity. See also: critical angle; dispersion; graded index profile; index profile; mode volume; multimode optical waveguide; normalized frequency; optical waveguide; refractive index; total internal reflection.

stimulated emission—Radiation emitted when the internal energy of a quantum mechanical system *drops* from an excited level to a lower level is induced by the simultaneous presence of radiant energy at the same frequency. An example is the radiation from an injection laser diode above lasing threshold.

superradiance—Amplification of spontaneously emitted radiation in a gain medium, characterized by moderate line narrowing and moderate directionality.

Note: This process is generally distinquished from lasing action by the absence of positive feedback and hence the absence of well-defined modes of oscillation.

tapered fiber waveguide—A waveguide that is tapered along its length; i.e., one whose transverse dimensions vary monotonically with length. Synonym: tapered transmission line.

tee coupler—A passive coupler that connects three ports. See also star coupler.

TE mode—Abbreviation for transverse electric mode. The Electric Vector is normal to the length of the waveguide and direction of wave propagation.

temporal coherence—See coherent.

temporally coherent radiation—See coherent.

thin film waveguide—An optically transparent dielectric film that when bounded by lower index material, forms a core that guides light. See also: optical waveguide.

threshold current—The driving current corresponding to lasing threshold. See also lasing threshold.

time coherence—See coherent.

TM mode—Abbreviation for transverse magnetic mode.

tolerance fields—A method of specifying dimensional tolerances for an optical waveguide core and cladding, utilizing four circles, concentric about the axis of the waveguide. The two smallest circles, having diameters $d \pm \Delta d$, inscribe and circumscribe the core-cladding interface. The two largest circles, having diameters $D \pm \Delta d$, inscribe and circumscribe the outer surface of the cladding.

total internal reflection—The reflection that occurs when light strikes an interface at incident angles (with respect to the normal) greater than the critical (Brewster) angle.

transducer—A device that converts one form of energy (optical, electrical, thermal, or mechanical) into another.

transfer function—A complex function (magnitude and phase) equal to the ratio of output to input as a function of modulation frequency. Synonym: baseband response function. Also: impulse response.

transmission loss—Total loss encountered in transmission through a system. See also: attenuation; optical density; reflection loss; transmittance.

transmission window—Synonym for spectral window.

transmissivity—The transmittance of a unit length of material, at a given wavelength, excluding the reflectance of the surfaces of the material; the intrinsic transmittance of the material, irrespective of the other parameters such as the reflectances of the surfaces (no longer in common usage).

transmittance—The ratio of transmitted power to incident power.
Note: In optics, frequently expressed as optical density or a percent; in communications applications, generally expressed in dB. Formerly "Transmission."

transverse electric (TE) mode—A mode whose electric field vector is normal (perpendicular) to the direction of propagation.
Note: In a planar dielectric waveguide (as within an injection laser diode), the field direction is parallel to the core-cladding interface and normal to the direction of propagation. In an optical waveguide, TE and TM Modes correspond to meridional rays.

transverse magnetic (TM) mode—A mode whose electric field vector is normal to the direction of propagation. (Thus the magnetic field vector is also normal. See Poynting Vector).
Note: In a planar dielectric waveguide (as within an injection laser diode), the field direction is parallel to the core-cladding interface. In an optical waveguide, TE and TM Modes correspond to meridonal rays.

transverse offset loss—Synonym for lateral offset loss.

ultraviolet (UV)—The region of the electromagnetic spectrum between the short-wavelength extreme of the visible spectrum (about 0.4 μm and 0.04 μm). In optical waveguide communications, the nominal wavelength region may be considered to be between 0.2 μm and 0.4 μm.

unbound mode—A mode whose power is predominently outside the core of the waveguide. Synonym: radiative mode.

VAD—Abbreviation for vapor phase axial deposition.

V number—Synonym for normalized frequency.

vapor phase axial deposition (VAD) technique—A method of fabricating an optical waveguide *preform* by forming small glass particles and depositing the particles on the end of a rod.

visible spectrum—See light.

wavefront—A continuous surface that is a locus of points having the same phase at a given instant.

waveguide dispersion—For each mode in an optical waveguide, that portion of total dispersion attributable to the dependence of the phase and group velocities on the geometric properties of the waveguide in particular, for circular waveguides, on ratio (a/λ), where (a) is core radius and (λ) is wavelength.

waveguide scattering—Scattering (other than material scattering) that is attributable to waveguide design and fabrication.

wavelength division multiplexing (wdm)—The provision of two or more channels over a common optical waveguide, the channels being differentiated by wavelength.

weakly guiding optical waveguide—A fiber waveguide for which the difference in refractive index between the core and cladding is small (usually less than 1.3).

y coupler—See tee coupler.

Index

Index

Other Bestsellers From TAB